2nd Edition

Good Tuning: A Pocket Guide

2nd Edition

By Gregory K. McMillan

ISA–The Instrumentation, Systems, and Automation Society

Copyright © 2005

ISA–The Instrumentation, Systems and Automation Society
67 Alexander Drive
P.O. Box 12277
Research Triangle Park, NC 27709

All rights reserved.

Printed in the United States of America.
10 9 8 7 6 5 4 3 2

ISBN   1-55617-940-5

No part of this work may be reproduced, stored in a retrieval system, or transmitted in any form or by any means, electronic, mechanical, photocopying, recording or otherwise, without the prior written permission of the publisher.

Library of Congress Cataloging-in-Publication Data in Process

McMillan, Gregory K., 1946-
  Pocket guide to good tuning / by Gregory K. McMillan.
    p. cm.
  ISBN 1-55617-940-5
  1. Automatic control--Handbooks, manuals, etc. I. Title.

ISA wishes to acknowledge the cooperation of those manufacturers, suppliers, and publishers who granted permission to reproduce material herein. The Society regrets any omission of credit that may have occurred and will make such corrections in future editions.

Notice

The information presented in this publication is for the general education of the reader. Because neither the author nor the publisher has any control over the use of the information by the reader, both the author and the publisher disclaim any and all liability of any kind arising out of such use. The reader is expected to exercise sound professional judgment in using any of the information presented in a particular application.

Additionally, neither the author nor the publisher have investigated or considered the effect of any patents on the ability of the reader to use any of the information in a particular application. The reader is responsible for reviewing any possible patents that may affect any particular use of the information presented.

Any references to commercial products in the work are cited as examples only. Neither the author nor the publisher endorses any referenced commercial product. Any trademarks or tradenames referenced belong to the respective owner of the mark or name. Neither the author nor the publisher makes any representation regarding the availability of any referenced commercial product at any time. The manufacturer's instructions on use of any commercial product must be followed at all times, even if in conflict with the information in this publication.

Table of Contents

Chapter 1.0—Best of the Basics 1

    1.1 Introduction 1
    1.2 Actions Speak Louder than Words 2
    1.3 Controller à la Mode 6
    1.4 Is That Your Final Response? 7
    1.5 The Key to Happiness 30
    1.6 Nothing Ventured, Nothing Gained 34
    1.7 Process Control as Taught versus as Practiced 42

Chapter 2.0—Tuning Settings and Methods 43

    2.1 First Ask the Operator 43
    2.2 Default and Typical Settings 43
    2.3 The General-purpose Closed-loop Tuning Method 47
    2.4 The Shortcut Open-loop Method 53
    2.5 Simplified Lambda Tuning 58
    2.6 Set Point Response and Load Rejection Capability 62

Chapter 3.0—Measurements and Valves 63

    3.1 Watch Out for Bad Actors 63

    3.2 Deadly Dead Band 64

    3.3 Sticky Situations 74

    3.4 Fouled Sensors 78

    3.5 Noisy Measurements 81

Chapter 4.0—Control Considerations.................... 85

    4.1 Auto Tuners............................... 85

    4.2 Uncommonly Good Practices for Common Loops 85

    4.3 Dead-Time Compensation and Warp Drive 88

    4.4 I Have So Much Feedforward, I Eat before I Am Hungry 92

    4.5 Cascade Control Tuning............. 95

    4.6 Keep the Secondary Loop on the Move 98

Chapter 5.0—Troubleshooting 99

    5.1 Patience, Heck, I Need to Solve the Problem 99

    5.2 Great Expectations and Practical Limitations................. 107

Chapter 6.0—Tuning Requirements for Various Applications 113

    6.1 Batch Control 115

    6.2 Blending 118

    6.3 Boilers 119

    6.4 Coils and Jackets 125

    6.5 Compressors 128

6.6 Crystallizers	130
6.7 Distillation Columns	132
6.8 Dryers	135
6.9 Evaporators	137
6.10 Extruders	138
6.11 Fermentors	140
6.12 Heat Exchangers	142
6.13 Neutralizers	143
6.14 Reactors	145
6.15 Remote Cascade	146
6.16 Sheets and Webs	147

Chapter 7.0—Adaptive Control 151

7.1 Learning the Terrain	151
7.2 Watching but Not Waiting	152
7.3 Shifting into High Gear	154
7.4 Back to the Future	156

Appendix A—Technical Terms in Process Control that Are Used Interchangeably 159

Appendix B—For Math Lovers Only 161

Appendix C—An Integral Part of Tuning 165

Appendix D—Closed Loop Time Constant 169

Index 171

1.0 – Best of the Basics

1.1 Introduction

Welcome to the wonderful world of proportional-integral-derivative (PID) controllers. This guide will cover the key points of good tuning and provide more than seventy rules of thumb. First, let's blow away some myths:

Myth 1 – It is always best to use one controller tuning method. False; the diversity of processes, control valves, control algorithms, and objectives makes this impractical.

Myth 2 - Controller tuning settings can be computed precisely. Not so. The variability and nonlinearity in nearly all processes and control valves makes this implausible. Any effort to get much more than one significant digit is questionable because any match to the plant is momentary. If you run a test for an auto tuner or manually compute settings ten times, you should expect ten different answers.

Roughly 75 percent of process control loops cause more variability running in the automatic mode than they do in the manual mode. A third of them oscillate as a result of nonlinearities such as valve dead band. Another third oscillate because of poor controller tuning. The remaining loops oscillate because of deficiencies in the control strategy. A well-designed control loop with proper tuning and a responsive control valve can minimize this variability. Because this means you can operate closer to constraints, good tuning can translate into increased production and profitability.

1.2 Actions Speak Louder than Words

The very first settings that must be right are the controller and valve actions. If these actions are not right, nothing else matters. The controller output will run off scale in the wrong direction regardless of the tuning settings.

The controller action sets the direction of a change in controller output from its proportional mode for every change in the controller's process variable (feedback measurement). If you choose *direct* action, an *increase* in process variable (PV) measurement will cause an *increase* in controller output that is proportional to its gain setting. Since the controller action must be the opposite of process action to provide feedback correction, you should use a direct-acting controller for a reverse-acting process except as noted later in this guide. Correspondingly, you should select *reverse* control action for a *direct*-acting process so an *increase* in process variable measurement will cause a *decrease* in controller output that is proportional to its gain setting, except as noted later. A *direct*-acting process is one in which the direction of the change in the process variable is the same as the direction of the change in the manipulated variable. A reverse-acting process is one in which the direction of the change in the process variable is opposite the direction of the change in the manipulated variable. The manipulated variable is most frequently the flow through a control valve, but it can also be the set point of a slave loop for a cascade control system or variable speed drive.

The valve action sets the display. For example, it determines whether a 100 percent output signal corresponds to a wide open or a fully closed valve. It also determines the direction of a change in the actual signal to the control valve when there is a change in the controller's output. In some analog controllers developed in the 1970s, such as the Fisher AC^2, the valve action affected only the display, not the actual signal. To compensate for this lack of signal reversal for a reverse-acting valve (i.e., an increase-to-close or fail-open valve), the control action had to be the opposite of the action that would normally be appropriate based on process action alone. Fortunately, the valve action corrects both the display and the actual valve signal in modern controllers, so the control action can be based solely on process action. However, the user should verify this before commissioning any loops. In control systems that use fieldbus blocks, the valve action should be set in the analog output (AO) block rather than in the PID controller block. This ensures that the "back-calculate" feature is operational for any function blocks (split range, characterization, and signal selection) that are connected between the PID and AO blocks. The signal can also be reversed in the current-to-pneumatic transducer (I/P) or in the positioner for a control valve. Before the advent of the smart positioner, it was preferable for the sake of visibility and maintainability that any reversal be done in the control room rather than at the valve. It is important to standardize on the location of the signal reversal to ensure that it is done and done only once. Table 1 summarizes how the controller action depends upon both the process and valve actions and on the signal reversal.

Table 1 – Controller Action

Process Action	Valve Action	Signal Reversal	Controller Action
Direct	Increase-Open	No	Reverse
Reverse	Increase-Open	No	Direct
Direct	Increase-Close	Yes	Reverse
Reverse	Increase-Close	Yes	Direct
Direct	Increase-Close	No	Direct
Reverse	Increase-Close	No	Reverse

Which brings us to rule of thumb number one.

Rule 1 – **The controller action should be the opposite of the process action unless there is an increase-to-close (fail-open) control valve for which there is no reversal of the valve signal.** This means that you should use reverse and direct-acting controllers for direct and reverse-acting processes,

respectively. The valve signal can be reversed for a fail-open valve at many places, but it is best done in the AO block of the control system.

1.3 Controller à la Mode

The names for the operational modes of the PID vary from manufacturer to manufacturer. Thus, the FOUNDATION™ Fieldbus modes listed next provide a uniformity that can be appreciated by all.

- **Auto** (automatic) – The operator locally sets the set point. PID action is active (closed loop). In older systems, this mode is also known as the *local mode*.

- **Cas** (cascade) – The set point comes from another loop. PID action is active (closed loop). This is also known in older systems as the remote mode or remote set point (*RSP*).

- **LO** (local override) – PID action is suspended. The controller output tracks an external signal to position the valve. This mode is typically used for auto tuning or to coordinate the loop with interlocks. In older systems, it is also known as *output tracking*.

- **Man** (manual) – The operator manually sets the output. PID action is suspended (open loop).

- **IMan** (initialization manual) – PID action is suspended because of an interruption in the forward path of the controller output. This is typically caused by a downstream block that is not in the cascade mode. The controller output is back-calculated to provide bumpless transfer.

- **RCas** (remote cascade) – The set point is remotely set, often by another computer. PID action is active (closed loop). This mode is also known in older systems as the *supervisory* mode.

- **ROut** (remote output) – The output is remotely set, often by a sequence or by another computer. PID action is suspended (open loop). In older systems, this mode is known as *direct digital control* (*DDC*).

1.4 Is That Your Final Response?

The contribution that the proportional action makes to the controller output is the error multiplied by the gain setting. The contribution made by the integral action is the integrated error multiplied by the reset and gain setting. The reset setting is repeats per minute and is the inverse of integral time or reset time (minutes per repeat). FOUNDATION™ Fieldbus will standardize the reset time setting as seconds per repeat and the rate (derivative) time setting as seconds. The contribution made by derivative mode is the rate of change of the error or process variable in percent (%), depending upon the type of algorithm, multiplied by the rate and gain settings.

When derivative action is on the process variable instead of on the control error, it works against a set point change. (The control error is the difference between the process variable and the set point.) The reason for this is that it doesn't know the process variable should be changing initially and that the brakes should only be applied to the process when it approaches set point. Using derivation action that is based on the change in control error will provide a faster initial takeoff and will suppress overshoot for a set point change. This is particularly advantageous for set points driven for batch control, advanced control, or cascade control. The improvement can translate into shorter cycle or transition times, an enhancement of the ability of slave loops to mitigate upsets, and less off-spec product because overshoot has been diminished.

When derivative action is used with a time constant there is a built-in filter that is about one-eighth (1/8) of the rate setting. However, you should use set point velocity limits to prevent a jolt to the output when there is a large step change in error from a manually entered set point. This is particularly important when you are using large gains or derivative action based on control error.

The PID algorithm uses percentage (%) input and output signals rather than engineering units. Thus, if you double the scale span of the input (error or process variable), you effectively halve the PID action. Correspondingly, if you double the scale span of the output (manipulated variable), you double the PID action. In fieldbus blocks, both input and output signals can be scaled with engineering units. Using output signal scaling will facilitate the manipulation of the slave loop's set point for cascade control. For controllers that use proportional band, you need to divide the proportional band into 100 percent to get the equivalent controller gain. Proportional band is the percentage change in error needed to cause a 100 percent change in output.

Proportional mode is expressed by the following equation (note that adjustments are gain or proportional band):

$$P_n = K_c * E_n$$

When the derivative mode acts on the process variable in percent (%PV), the equation becomes as follows for the series form:

$$P_n = K_c * D_n$$

Integral mode is expressed by the following equation (note that adjustments are integral time or reset):

$$I_n = K_c * 1/T_i * (E_n * T_s) + I_{n-1}$$

When the derivative mode acts on %PV, the equation becomes as follows:

$$I_n = K_c * 1/T_i * (D_n * T_s) + I_{n-1}$$

The equation for derivative mode is as follows (adjustments are derivative time or rate):

$$D_n = K_c * T_d * (E_n - E_{n-1})/ T_s$$

The series form derivative on %PV:

$$D_n = T_d * (\%PV_n - \%PV_{n-1}) / T_s$$

Note the inverse relationships across controllers!

$$\text{Gain } (K_c) = 100\% / PB$$

Reset action (repeats/minute) = 60 / T_i

Where:

E_n = error at scan n (%)
K_c = controller gain (dimensionless)
PB = controller proportional band (%)
P_n = contribution of the proportional mode at scan n (%)
$\%PV_n$ = process variable at scan n (%)
I_n = contribution of the integral mode at scan n (%)
D_n = contribution of the derivative mode at scan n (%)
T_d = derivative time or rate time setting (seconds)
T_i = integral time or reset time setting (seconds/repeat)
T_s = scan time or update time of PID controller (seconds)

Rule 2 - **If you halve the scale span of a controlled (process) variable or double the span of a manipulated variable (i.e., set point scale or linear valve size), you need to halve the controller gain to get the same PID action.** For controlled variables, the PID gain is proportional to the measurement

calibration span. Often these spans are narrowed since accuracy is a percentage of span. For control valves, the PID gain setting is inversely proportional to the slope of the installed valve characteristic at your operating point. If a valve is sized too small or too large, the operating point ends up on the flat portion of the installed valve characteristic curve. For butterfly valves, the curve gets excessively flat below 15 percent and above 55 percent of valve position.

Figure 1 shows the combined response of the PID controller modes to a step change in the process variable (%PV). The proportional mode provides a step change in the controller output ($\Delta\%CO_1$). If there is no further change in the %PV, there is no additional change in the output even though there is a persistent error (offset). The size of the offset is inversely proportional to the controller gain. Integral action will ramp the output unless the error is zero. Since the error is hardly ever exactly zero, reset is always driving the output. The contribution made by the integral mode will equal the contribution made by the proportional mode in the integral time ($\Delta\%CO_2 = \Delta\%CO_1$). Hence, the integral time setting is the time it takes to repeat the proportional contribution (seconds per repeat). The contribution made by the derivative mode for a step change is a hump because

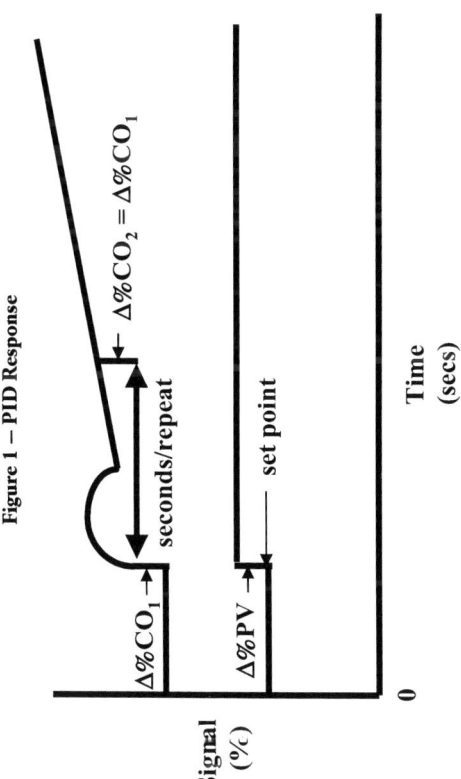

of the built-in filter, which is about one-eighth (1/8) of the rate setting. Otherwise, there would be a spike in the output.

If the temperature is below set point for a reactor, as shown in Figure 2, should the steam or water valve be open? After looking at a faceplate or the digital value for temperature on a graphic display, most people think the steam valve is open when the temperature is below set point. Reset provides a direction of action that is consistent with human expectation. However, the proper direction for a change in controller output and the split-ranged control valve depends upon the trajectory of the process variable (PV). If the temperature is rapidly and sharply increasing, the coolant valve should be opening. Gain and rate action will recognize that a set point is being approached and position the valves correctly to prevent overshoot. In contrast, reset has no sense of direction and sacrifices future results for immediate satisfaction. Most reactors, evaporators, crystallizers, and columns in the process industry have too much reset. On two separate applications in chemical plants, for example, it was reported that the controller was seriously malfunctioning because the wrong valve was supposedly open, when in reality it was just gain and rate doing their job to prevent overshoot.

Figure 2 – Reset Gives Them What They Want

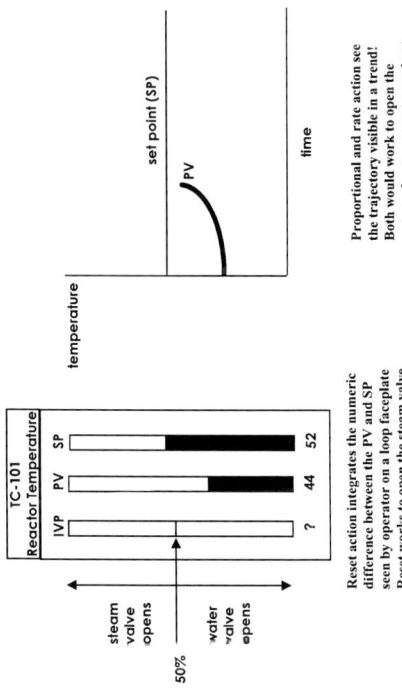

Reset action integrates the numeric difference between the PV and SP seen by operator on a loop faceplate Reset works to open the steam valve (reset has no sense of direction).

Proportional and rate action see the trajectory visible in a trend! Both would work to open the water valve to prevent overshoot.

BEST OF THE BASICS

Figures 3a and 3b show the effect of gain setting on a set point response. If the gain is too small, the approach to set point is too slow. If the gain is too large, the response will develop oscillations. If the dead time is much larger than the time constant and the gain action is too large compared to the reset action, the response will momentarily flatten out (falter or hesitate) well below the set point. Loops that are dead-time dominant tend to have too much gain and not enough reset action. For nearly all other loops in the process industry, the opposite is true. The size of the dead time relative to the time constant and the degree of self-regulation (ability to reach a steady state when the loop is in manual) determine which tuning methods you should employ.

It is particularly important to maximize the gain so you can achieve tight control of loops that can ramp away from set point, such as gas pressure. Maximizing the gain is also important for loops that can run away from set point, such as an exothermic reactor temperature. Finally, gain must be maximized to speed up the set point response for advanced control, batch control, and cascade control as well as for startup sequences. However, gain readily passes variability in the process variable to the output, which can increase overall variability and interaction.

Figure 3a – Effect of Gain Setting on Set Point Response for a Time Constant Much Larger than Dead Time

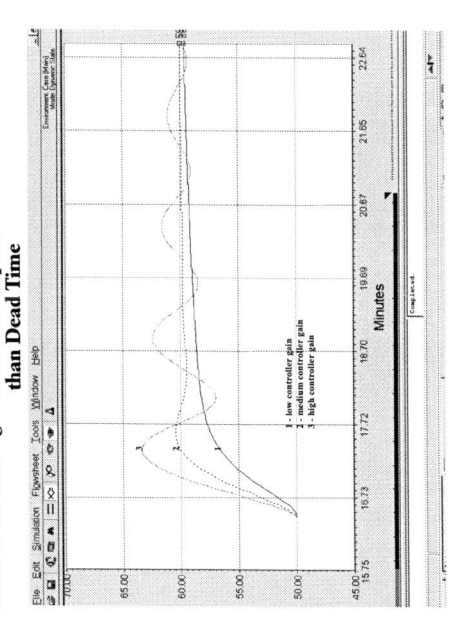

Figure 3b – Effect of Gain Setting on Set Point Response for a Dead Time Much Larger than Time Constant (Dead Time Dominant)

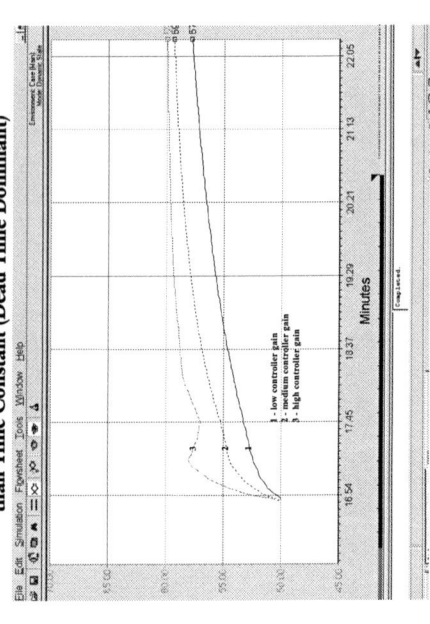

You must set the process variable filter to ensure that fluctuations from measurement noise stay within the dead band of the control valve. The variability amplification by gain can be particularly troublesome if there is inadequate smoothing because of insufficient volumes or mixing. These loops tend to be dead-time dominant, and they exhibit abrupt responses (i.e., no smoothing by a time constant). Thus, you should reduce the gain setting for loops on pipelines, desuperheaters, plug flow reactors, heat exchangers, static mixers, conveyors, spinning (fibers), and sheets (webs).

Rule 3 – **Increase the gain to achieve tight level, stirred reactor, or column control and to speed up the set point response for advanced, batch, cascade, and sequence control.** For high gains it is especially important that you set a process variable filter to ensure that fluctuations in the PID output are smaller than the valve dead band. You should also establish set point velocity limits so the output doesn't jerk when the operator changes the set point.

Rule 4 – **Decrease the gain to provide a smoother, slower, and more stable response; reduce interaction between loops; and reduce the amplification of variability.** Pulp and paper, fiber, and sheet processes use less gain (higher proportional band) and more reset action (smaller reset time). They also are much more sensitive to variability from valve dead band. The dead time is larger than the process time constant for most of these loops.

Rule 5 – **For level control of surge tanks, use a gain, such as 1.5, that is just large enough to avoid actuating the alarms set to help prevent the tank from overflowing or running dry.** You can estimate the gain as follows: it is the maximum needed change in controller output divided by the difference between the high- and low-level alarms. For example, if a 60 percent change in controller output will always keep the level between 30 percent and 70 percent, the correct controller gain is 1.5. You can add an error-squared algorithm, where gain is proportional to error, to attenuate control action near the set point. This is a generally effective way to reduce the effect of noise for all types of level control loops on the manipulated flow.

 *Rule 6* - **For residence time control and for material balance control, use as high a gain for level control as possible, such as 5.0, that does not wear out the valves or upset other loops.** Continuous reactors, evaporators, crystallizers, and thermosyphon reboilers often need to maintain a level accurately, particularly if they are being pushed beyond nameplate production rates. Level control also needs to be tight when it manipulates a reflux flow for a column or a reactant makeup flow for a recycle tank.

The Ziegler-Nichols (J. G. Ziegler and N. B. Nichols, "Optimum Settings for Automatic Controllers," *Transactions of the ASME*, vol. 64, Nov. 1942, p. 759) tuning method depicts a goal of quarter-amplitude response where each subsequent peak in an oscillatory response of the PID controller is one-fourth of the previous peak. This goal will generally provide a minimum peak error, but the controller is too close to the edge of instability. An increase of just 25 percent in the process gain or dead time can cause severe oscillations. By simply cutting the controller gain in half, you will make Ziegler-Nichols tuning sufficiently robust. You will also make it look more like Internal Model Control (IMC) tuning for stirred reactors, evaporators, crystallizers, and columns. If the controller manipulates a control valve instead of a flow set point, you need to reduce the gain for the steepest part of the installed valve characteristic curve.

For loops that will reach a steady state when put in manual, decreasing the gain will always improve their stability. For exothermic reactors where a runaway condition can develop, there is a window of allowable gains in which instability can be caused by a gain that is too small or too large. For level loops with reset action, there is also a window in which a gain that is too small can cause slow, nearly sustained oscillations.

Rule 7 – **If you have a quarter-amplitude response, you should decrease the gain to promote stability.** Generally, halving the controller gain is sufficient. This assumes that the reset action is not excessive (i.e., integral or reset time is too small). If the manipulated variable is a control valve and the operating point moves from the flat to the steep portion of the installed characteristic, you may need to cut the gain by a factor of three or four.

Figure 4 shows the effect of reset setting on a set point response. If the reset action is too small (i.e., the reset time too large), it takes a long time

Figure 4 – Effect of Reset Setting on Set Point Response

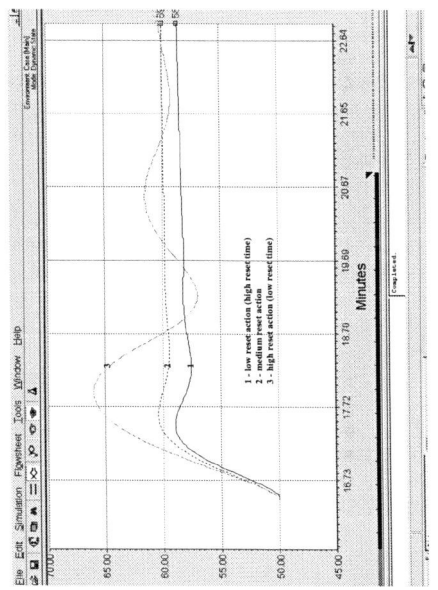

1 – low reset action (high reset time)
2 – medium reset action
3 – high reset action (low reset time)

to eliminate the remaining error or offset. Whereas insufficient gain slows down the initial approach, inadequate reset action slows down the final approach to set point. If the reset action is too large (reset time is too small), there is excessive overshoot. The addition of reset action reduces controller stability and is dangerous for exothermic reactor temperature loops. Polymerization reactors often use proportional-plus-derivative temperature controllers (i.e., no integral action). For level loops, using a reset time of smaller than 3,000 seconds per repeat makes it necessary to use relatively high gains (5.0 for large volumes) to prevent slow, nearly sustained oscillations.

Both reset and gain can cause oscillations, and a high gain setting can aggravate an overshoot caused by reset action. For this reason, checking the ratio of the reset time to the oscillation period can help you distinguish the main culprit. You will need to decrease the reset action (i.e., increase the reset time) of the loop if the ratio is much less than 0.5 for vessel or column temperature, level, and gas pressure control or if it is much less than 0.1 for flow, liquid pressure, pipeline, spin-line, conveyor, or sheet (web) control.

Rule 8 – **If there is excessive overshoot and oscillation, decrease the reset action (increase the reset time).** If the current reset time setting is less than half the oscillation period for loops on mixed volumes or less than a tenth of the period for any loop, there is too much reset action. If the oscillation period changes by 25 percent or more as you change reset action, it is a sign that the oscillation is caused or aggravated by reset action.

Rule 9 – **For level loops with low controller gains (< 5.0), reset action must be greatly restricted (i.e., reset time increased to 50 or more minutes) to help mitigate nearly sustained oscillations.** Level loops are the opposite of most other loops in that you can increase the reset action (decrease the reset time) as you increase the gain (see page 109 for more detail).

Figure 5 shows the effect of rate setting on a set point response. A rate setting that is too small increases any overshoot caused by reset action. A setting that is too large causes the approach to set point to staircase and if large enough it can cause fast oscillations.

Figure 5 – Effect of Rate Setting on Set Point Response

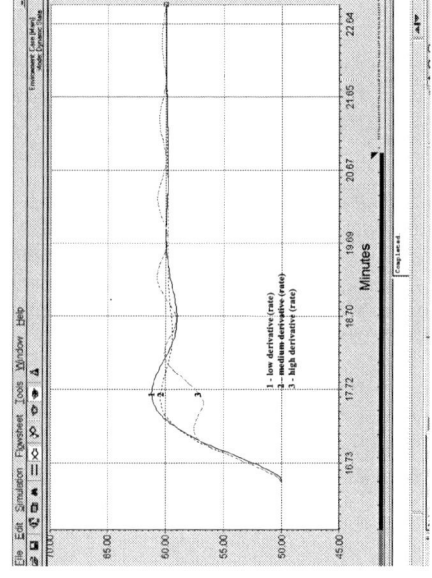

Rate can be used whenever there is a smooth and slow open-loop process response. Rate cannot be used where there is excessive noise, chatter, inverse response, or an abrupt response. It can be used primarily on temperature loops that manipulate heating or cooling where there is a temperature transmitter with a narrow calibration span (large spans cause analog/digital [A/D] chatter). Analyzers that have a cycle time (sample and hold) have signals that are too noisy and abrupt for derivative action. Continuous concentration measurements made by electrodes or inline devices (such as capacitance, conductivity, density, microwave, mass spectrometry, pH, and viscosity for mixed volumes) may allow you to use derivative mode if you add a filter to attenuate the high-frequency measurement noise.

If the measurement is continuous but is used for pipeline control, the control response is too abrupt. This is because the loop dead time is large compared to the largest time constant.

There is noise in every process variable. If the noise is greater than the measurement resolution limit, the loop sees it. If it causes an output change greater than the final element or control valve resolution limit, loop variability will be worse in auto than in manual, particularly if there is any rate action. While it is true that increasing the measurement and valve resolution limit reduces the loop's sensitivity to noise, it also degrades its ability to recover from load upsets. If the measurement doesn't see or correct an upset because the control error is smaller than a resolution limit, the control loop performance deteriorates.

Poor control valve resolution is a problem for any loop. If the valve doesn't move, there is no correction. On the other hand, poor measurement resolution is primarily a problem for PID temperature loops. The large-span ranges found in computer input cards or transmitter calibration can result in a resolution limit of 0.25°F or more. It is impractical to expect control that is tighter than twice the resolution limit or 0.5°F. Also, for a change of 0.25°F per minute, an additional minute of dead time is introduced into the loop. Short scan times also introduce chatter because the actual temperature change is small compared to the noise (i.e., the signal-to-noise ratio is too small). The scan time for most temperature loops is too short. Poor measurement resolution and fast scan times are the main reasons why rate cannot be used in temperature loops. Model Predictive Controllers (MPC) are much less sensitive to measurement resolution because the scan time is large and the move in the manipulated variable is based on the complete trajectory and not the current measurement like the PID controller. Poor valve resolution is as much a problem for the MPC as for the PID, unless you set the minimum size of a move to exceed the resolution limit. The resolution limit is unfortunately a function of direction and position.

Rule 10 – **Rate is primarily used in temperature loops that have narrow span transmitters and slow scan times.** However, it is also used for continuous analytical measurements such as capacitance, conductivity, density, mass spectrometry, microwave, pH, and viscosity for concentration control of mixed volumes. If the temperature control is not intended for a mixed volume, the rate setting is very small (6 seconds) and is mostly used to compensate for the thermowell lag. In every case, you must set the measurement filter large enough to keep output fluctuations within the valve dead band.

Rule 11 – **Most temperature and level loops chatter because the PID scan time is too small.** The true temperature or level change within the scan time must be large compared to the A/D resolution limit (0.05%); otherwise, the signal-to-noise ratio is too small.

1.5 The Key to Happiness

When I was four years old and sitting on my daddy's knee, he said, "Son, I have just one thing to say to you—dead time." Well, it took me forty years to appreciate the significance of his words of wisdom. If a loop has no dead time or noise, perfect control is possible and an infinite gain permissible. There is no tuning issue. Without dead time, I would be out of a job.

Dead time is that period of time from the start of a disturbance until the controller makes a correction that arrives at the same point in the loop at which the disturbance entered. The controller needs to see the upset, react to it, and get the correction to the right place. To appreciate dead time imagine that you go to a party and start drinking. The period of time between the first drink and when you eventually bypass the next round is dead time.

While zero dead time is not possible, a decrease in dead time reduces the effect of nonlinearities, such as changes in the time constants and steady-state gains of the loop, and makes the loop easier to tune. It is desirable that the largest loop time constant be in the process because it will slow down the divergence of the measurement from the set point during an upset and give the controller a chance to catch up. All other time constants are undesirable and create additional dead time. The major time constant can be approximated as follows: it is the time it takes to reach about 63 percent of the final value after the dead time. It takes one dead time and four time constants for a response to reach 98 percent of its final value.

The open-loop steady-state gain is the percentage change in the process variable for a percentage change in output after all transients have died out. The controller is in manual. A high steady-state gain is both a curse and blessing. It can improve the control of the true process variable by making the measurement more sensitive. This is particularly important for inferring composition from temperature or concentration from pH. However, a high steady-state gain can make it more difficult--and sometimes impossible--to control the measured variable. For example, excessive oscillation can result from a small amount of stick and slip in a reflux valve for an acrylonitrile and water distillation or in a reagent valve for sulfuric acid and sodium hydroxide neutralization. A low steady-state gain often results when a throttle position is on the excessively flat portion of the installed valve characteristic. The process variable will wander and respond to the conditions of a related loop more than from the manipulated variable.

Figure 6 – Loop Block Diagram

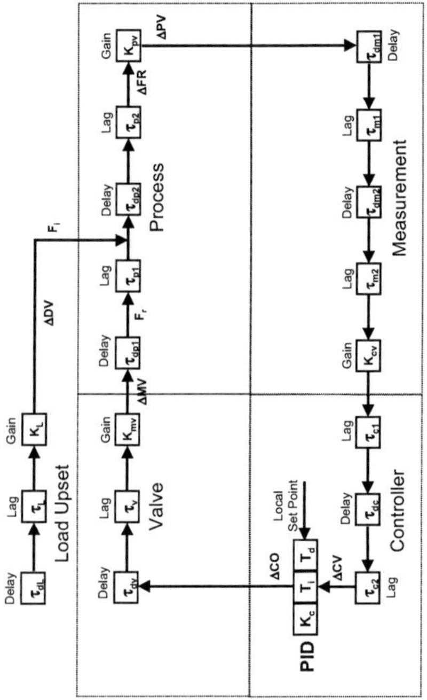

The block diagram in Figure 6 shows all the dead times, time constants, and steady-state gains in a control loop. Feedback control corresponds to the signal making one complete circle around the block diagram. The total dead time is approximately the sum of all the pure dead times and small time constants as you traverse the loop. The dead time from valve dead band is inversely proportional to the rate of change of the controller output, and the dead time from transportation delays is inversely proportional to throughput. The diagram in Figure 6 has many uses--including preventing guests from overstaying their welcome. Just show this slide and start talking about dead time and you will be amazed at how quickly the place empties. There is a safety issue, however, in that guests can get trampled in the rush for the door.

The following list summarizes the three key variables and their relationship to loop performance:

1. *Dead Time or Time Delay (τ_d)*

    The most important of the three key loop variables

    Delays controller's ability to see and react to upset

    Perfect control is possible for zero dead time

Nonlinearities become less important as the dead time decreases

2. *Time Constant (τ)*

    Better control is possible for a large *process* time constant downstream of the load upsets

    Time constants in series create dead time

    Measurement time constants give the illusion of providing better control

3. *Steady-State Gain (K)*

    The valve and process steady-state gains are usually nonlinear

    High steady-state gain causes overreaction and oscillations

    Low steady-state gain causes loss of sensitivity and wandering

1.6 Nothing Ventured, Nothing Gained

There is a lot to be gained from having a better understanding of the overall open-loop gain (K_o). However, if you don't have time to study the relationships revealed by the equations presented in this section, you should skip ahead to Section 2.0 on Tuning Settings and Methods.

The controller gain is inversely proportional to the open-loop gain for all types of loops. For loops that are dead-time dominant (i.e., dead time is much greater than the largest time constant), the controller gain can be simply set as one-fourth of the inverse of the open-loop gain. For these same loops the reset time can be set as one-fourth of the dead time. The result is a smooth stable response similar to what you would get from Lambda tuning, which will be discussed (see Section 2.5) as one of the preferred tuning methods for dead-time dominant loops. For loops with a healthy time constant, the controller gain is also proportional to the ratio of the time constant to the dead time. For these loops, the smooth response that is provided by a large time constant (i.e., residence time) of a mixed volume enables you to use a higher controller gain for tighter concentration, gas pressure, and temperature control.

If the loop time constant is much larger than the dead time ($\tau/\tau_d \gg 1$), then the controller gain is proportional to this ratio besides the inverse of the open loop again:

$$K_c \cong 0.25 * \tau / (\tau_d * K_o)$$

If the loop is dead-time dominant ($\tau/\tau_d \ll 1$), then the controller gain is proportional to just the inverse of the open loop gain:

$$K_c = 0.25 * 1 / K_o$$

The manipulated variable gain is linear for a variable speed drive (VSD). For a control valve it is the slope of the installed characteristic curve and is typically very nonlinear. For operating points on the upper part of the installed characteristic curve of a butterfly valve, the slope can be so flat that the valve gain approaches zero.

Note that for concentration and pressure loops, the process variable should be plotted against a flow ratio (e.g., column temperature versus distillate-to-feed ratio, heat exchanger temperature versus coolant-to-feed ratio, and pH versus reagent-to-feed ratio). The process gain is the slope of this curve and is highly nonlinear. At high flow ratios, the slope can be so flat that the process gain approaches zero. The open-loop gain has a flow ratio gain that is inversely proportional to the feed flow, which is an additional source of nonlinearity for pipeline, desuperheater, static mixer, and exchanger temperature or concentration control. In these applications, an equal-percentage valve characteristic is desirable since the valve gain is proportional to flow and cancels out the flow ratio gain. However, for loops on agitated vessels and columns the controller gain is proportional to the process time constant (residence time), which is inversely proportional to feed flow. Since, as we mentioned previously, the controller gain is also inversely proportional to the open-loop gain, the introduction of an equal-percentage valve characteristic is bad news because the flow ratio gain was cancelled out by the process time constant for the controller gain.

The process gain for flow loops is 1. For pressure loops downstream of a compressor, fan, or pump, the process gain is the slope at the operating point on the characteristic curve of the compressor, fan, or pump. At low flows, the slope can be so flat that the process gain approaches zero.

The controlled variable gain is the inverse of the calibration span for the controlled variable. It is linear except when a square root extractor is not used on a differential head meter.

Rule 12 – For dead-time-dominant loops, set the gain equal to about one-fourth of the inverse of the open-loop gain and the reset time (seconds per repeat) equal to

about one-fourth of the dead time. Rate action is not used. This provides a smooth stable response similar to Lambda tuning.

Rule 13 – **Use equal-percentage valve trim for pipeline, desuperheater, static mixer, and heat exchanger control to make the loop more linear.** This assumes that the installed characteristic is close to the inherent equal-percentage trim characteristic.

For concentration and temperature loops, the open-loop gain is a steady-state gain that has a flow ratio gain K_{fr}, which is the inverse of the feed flow:

$$K_o = \frac{\Delta CV}{\Delta CO} = \frac{\Delta MV}{\Delta CO} \cdot \frac{\Delta FR}{\Delta MV} \cdot \frac{\Delta PV}{\Delta FR} \cdot \frac{\Delta CV}{\Delta PV}$$

$$K_o = K_{mv} * K_{fr} * K_{pv} * K_{cv}$$

$$K_{fr} = \frac{1}{F_f}$$

For flow loops, the open-loop gain is a steady-state gain that has a unity process gain and no flow ratio gain (K_{fr} omitted and $K_{pv} = 1$):

$$K_o = K_{mv} * K_{cv}$$

For pressure loops, the open-loop gain is a steady-state gain with a process gain K_{pv}, which is the slope of the pump or compressor curve, and no flow ratio gain (K_{fr} omitted and K_{pv} = curve slope):

$$K_o = K_{mv} * K_{pv} * K_{cv}$$

For level loops, the open-loop gain is a ramp rate and uses K_i instead of K_{pv} to denote an integrator gain. K_i is the inverse of the product of fluid density and the vessel's cross-sectional area. There is no flow ratio gain (K_{fr} omitted):

$$K_o = K_{mv} * K_i * K_{cv}$$

$$K_i = \frac{1}{\text{density} \bullet \text{area}}$$

Where:

F_f = feed flow (pph)
K_i = level loop integrator gain (ft/lb)
K_o = overall open-loop gain (1/hr for level, otherwise dimensionless)
K_{mv} = manipulated variable (valve or VSD) steady-state gain (pph/%)
K_{fr} = flow ratio steady-state gain (inverse of feed flow) (1/pph)

K_{pv} = process variable steady-state gain (wtfrac/%) (degC/%) (psi/%)

K_{cv} = controlled variable steady-state gain (%/wtfrac) (%/degC)(%/psi) (%/pph)(%/ft)

ΔCO = change in the controller output (%)

ΔCV = change in the controlled variable (%)

ΔPV = change in the process variable (wtfrac) (degC) (psi) (pph) (ft)

ΔFR = change in the flow ratio (pph/pph)

ΔMV = manipulated variable or valve flow (pph)

Figure 7 shows the dead time, time constant, and steady-state gain for a process that reaches a steady state (i.e., self-regulating) and for an integrator such as level. In both cases, the process variable response is to change the controller output with the loop in manual (open-loop response). Figure 7 also shows the equations used to approximate an integrator gain as equal to the initial ramp rate of a self-regulating response. This initial ramp rate is approximately the steady-state gain divided by the dominant time constant. Auto tuners that compute settings from an open-loop response use this approximation to convert an integrator into an equivalent

Figure 7 – Integrator Gain (Ramp Rate)

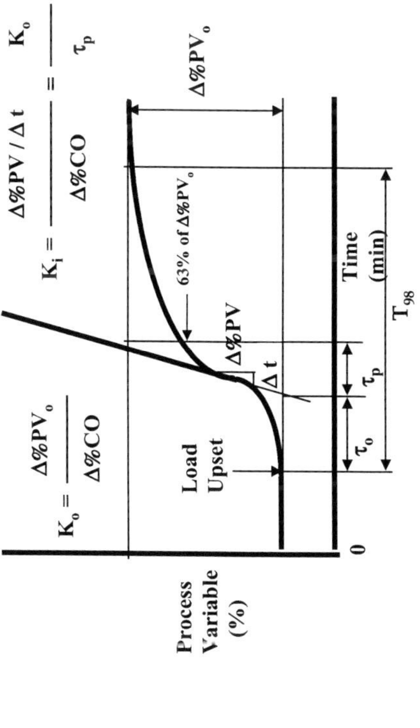

$$K_o = \frac{\Delta\%PV_o}{\Delta\%CO}$$

$$K_i = \frac{\Delta\%PV/\Delta t}{\Delta\%CO} = \frac{K_o}{\tau_p}$$

time constant and steady-state gain because they cannot handle integrator gain. This method works well when the dead time is small compared to the ramp rate. The dead time is the time required for the PV to get out of the noise band.

1.7 Process Control as Taught versus as Practiced

Most books, courses, and papers concentrate on a set point response and if a load is introduced, it is usually shown as an addition to the process output, which is really just measurement noise. In the process industry, the real load upsets, such as feed flow, temperature, and composition, enter as inputs to the process as shown in Figure 6. However, tuning studies that concentrate solely on the set point response have seriously compromised the load rejection capability for processes with a large process time constant. The optimum load response should reach a peak in just over one total loop dead time and show a return to set point that is just as fast. This is generally achieved by the controller initially overdriving its output past its final resting position. The load response can easily be checked by momentarily putting the controller in manual, making a step change in the controller output, and then returning the controller to automatic retaining the original set point.

2.0–Tuning Settings and Methods

2.1 First Ask the Operator

You should test new tuning settings by changing the controller set point and output in both directions and observing whether the response is smooth and fast enough for these set point changes and load upsets. Don't leave these new settings over night until you are sure they are right. Before entering new settings or trying tuning methods, ask the operator for permission to tune the loop. Also ask him or her what the maximum allowable size can be for each direction of a step change in the controller set point and output without causing a problem. You should put most analog controllers and some digital controllers in manual before entering a change in gain to avoid bumping the controller output.

2.2 Default and Typical Settings

Tables 2 and 3 show the default and typical tuning settings for controllers whose reset setting is in repeats per minute and seconds per repeat, respectively. The number in front of the parentheses in each column is a default setting, a number that can be used for a download or for a guess if nothing is known about the loop and no test can be performed. The numbers within the parentheses represent a range of typical values. If your tuning setting falls outside the suggested range, this doesn't necessarily mean the tuning setting is wrong. It does suggest that the loop is unusual or has some problems such as interaction, noise, a coated sensor, or a sticky control valve. The last column shows the tuning methods that generally give the best results. The "general-purpose" tuning method logic shown in Figure 8 is the closed-loop method (CLM), which is referenced in Tables 2 and 3. While this method is best for gas pressure, reactor, and level loops, it can be used for most other loops as well, particularly if the loop's set point is being driven for advanced, batch, cascade, or sequence control and there are no interaction issues.

Table 2 – Default and Typical PID Settings (scan in sec, reset in rep/min, and rate in minutes; λ = Lambda, CLM = Closed-loop method; SCM = Shortcut method)

Application Type	Scan (seconds)	Gain	Reset (repeats)	Rate (minutes)	Method
Liquid Flow/Press	1 (0.2-2)	0.3 (0.2-0.8)	10 (5-50)	0.0 (0.0-0.02)	λ
Tight Liquid Level	5 (1.0-30)	5.0 (0.5-25)*	0.1 (0.0-0.5)	0.0 (0.0-1.0)	CLM
Gas Pressure (psig)	0.2 (0.02-1)	5.0 (0.5-20)	0.2 (0.1-1.0)	0.05 (0.0-0.5)	CLM
Reactor pH	2 (1.0-5)	1.0 (0.001-50)	0.5 (0.1-1.0)	0.5 (0.1-2.0)	SCM
Neutralizer pH	2 (1.0-5)	0.1 (0.001-10)	0.2 (0.1-1.0)	1.2 (0.1-2.0)	SCM
Inline pH	1 (0.2-2)	0.2 (0.1-0.3)	2 (1-4)	0.0 (0.0-0.05)	λ
Reactor Temperature	5 (2.0-15)	5.0 (1.0-15)	0.2 (0.05-0.5)	1.2 (0.5-5.0)	CLM
Inline Temperature	2 (1.0-5)	0.5 (0.2-2.0)	1.0 (0.5-5.0)	0.2 (0.2-1.0)	λ
Column Temperature	10 (2.0-30)	0.5 (0.1-10)	0.2 (0.05-0.5)	1.2 (0.5-10)	SCM

* An error/square algorithm or gain scheduling should be used for level loops with gains < 5

Table 3 – Default and Typical PID Settings (scan in sec, reset in sec/rep, and rate in sec; λ = Lambda, CLM = Closed-loop method, SCM = Shortcut method)

Application Type	Scan (seconds)	Gain	Reset (seconds)	Rate (seconds)	Method
Liquid Flow/Press	1 (0.2-2)	0.3 (0.2-0.8)	6 (1-12)	0 (0-2)	λ
Tight Liquid Level	5 (1.0-30)	5.0 (0.5-25)*	600 (120-6000)	0 (0-60)	CLM
Gas Pressure (psig)	0.2 (0.02-1)	5.0 (0.5-20)	300 (60-600)	3 (0-30)	CLM
Reactor pH	2 (1.0-5)	1.0 (0.001-50)	120 (60-600)	30 (6-30)	SCM
Neutralizer pH	2 (1.0-5)	0.1 (0.001-10)	300 (60-600)	70 (6-120)	SCM
Inline pH	1 (0.2-2)	0.2 (0.1-0.3)	30 (15-60)	0 (0-3)	λ
Reactor Temperature	5 (2.0-15)	5.0 (1.0-15)	300 (300-3000)	70 (30-300)	CLM
Inline Temperature	2 (1.0-5)	0.5 (0.2-2.0)	60 (12-120)	12 (12-60)	λ
Column Temperature	10 (2.0-30)	0.5 (0.1-10)	300 (300-3000)	70 (30-600)	SCM

* An error/square algorithm or gain scheduling should be used for level loops with gains < 5

2.3 The General-purpose Closed-loop Tuning Method

A closed-loop (controller in auto) method has the following advantages over open-loop (controller in manual) methods:

1. It forces the user to find the maximum controller gain to minimize peak error and dead time from dead band. It also forces the user to find out whether reset and rate is even needed, and it ensures rapid set point response and gets inside the window of allowable controller gains for integrating and runaway loops.

2. Loops stay in automatic, which is safer for difficult or very fast loops.

3. It includes the effects of valve stick-slip and dead band.

4. It includes the dynamics and peculiar features of controller algorithms.

5. It includes nonlinearities that are dependent on direction and rate of change.

6. It facilitates the tuning of the master (outer) loop of a cascade system for an oscillating inner (slave) loop.

This closed-loop procedure uses rapidly decaying oscillations as shown in Figure 9 instead of the sustained oscillations proposed by the original Ziegler-Nichols method. The measured period will be larger for damped oscillations in industrial applications mostly because of valve dead band, but the error will be in the safe direction in terms of a larger, more stable reset time setting. The closed-loop procedure should also result in about half of the gain estimated by the Ziegler-Nichols method.

Rule 14 – **Use the general-purpose closed-loop method if the loop must stay in auto or if it is particularly important to maximize the gain for tight control and a fast set point response.** However, for dead-time-dominant loops, you should substantially decrease the factor for the reset time. This will prevent the set point response from faltering because of too much gain action and not enough reset action. For temperature and pressure loops on exothermic reactors, it is especially important to use this method to prevent a runaway.

Figure 8 – General-Purpose Closed-loop Method

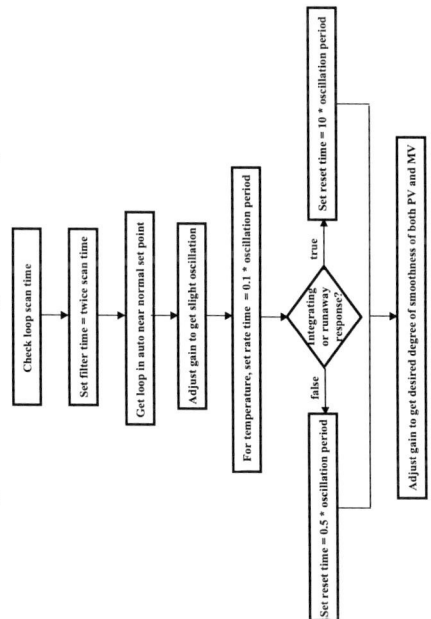

- Check loop scan time
- Set filter time = twice scan time
- Get loop in auto near normal set point
- Adjust gain to get slight oscillation
- For temperature, set rate time = 0.1 * oscillation period
- Integrating or runaway response?
  - false: Set reset time = 0.5 * oscillation period
  - true: Set reset time = 10 * oscillation period
- Adjust gain to get desired degree of smoothness of both PV and MV

TUNING SETTINGS AND METHODS

Figure 9 – Slight Oscillation

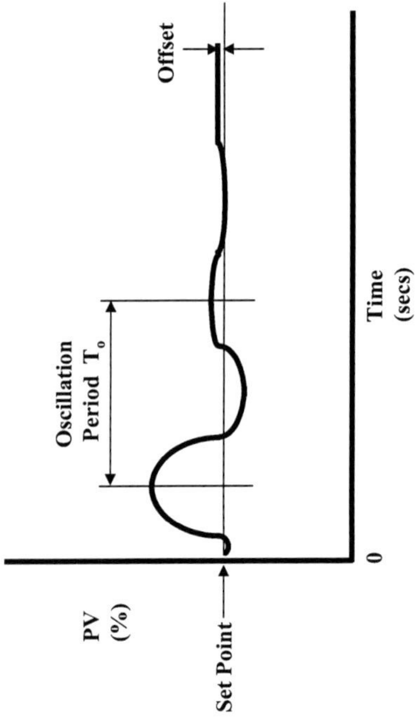

A list of steps for the "Closed Loop Method" is as follows:

1. Put the controller in automatic at normal set point. If it is important not to make big changes in the manipulated variable (controller output), narrow the controller output limits to restrict valve movement.

2. For level, gas pressure, and reactor loops decrease the reset action (i.e., increase the reset time) by a factor of ten if possible and trend-record the process variable (PV) and controller output (CO).

3. Add a PV filter to keep output fluctuations within the dead band of valve caused by noise.

4. Bump the controller set point and output and increase the controller gain if necessary to get a slight oscillation.

5. Stop increasing the gain when the loop *starts* to oscillate or the gain has reached your comfort limit. Then note the period. For gain settings greater than 1, the oscillation will be more recognizable in the controller output (CO). Make sure CO stays on scale within the valve's good throttle range.

6. Reduce the gain until the oscillation just disappears so recovery is smooth.

7. For a temperature loop with a smooth response (no chatter, inverse, square wave, or interaction), use rate action. If the gain is larger than 10, reset and rate action are not needed. If the manipulated flow will upset other loops, decrease the gain or use error-squared control. If a high gain is used, set a velocity limit for the set point and configure the set point so it tracks the PV in manual and ROUT (DDC). This will enable the loop to restart.

8. If rate action is used, set the rate time equal to one-tenth (1/10) of the period and set reset time equal to half the period. If rate action is not used, cut the gain by 50 percent. If the loop is clearly dead-time dominant, increase the reset action so the reset time is about one-eighth (1/8) of the loop period. Make another set point change and adjust the gain to get a smooth response. Do not become any more aggressive than a slight oscillation.

9. If you select gains smaller than 5 for level, decrease the reset action (i.e., increase the reset to more than 3,000 seconds), add feed-

forward, and add rate if there is no inverse response or interaction.

2.4 The Shortcut Open-loop Method

The shortcut method is ideally suited for very slow responses such as column temperature where there is a good control valve and positioner and you need to get a quick estimate of the controller tuning settings. This method looks at the change in ramp rate of the %PV as shown in Figure 10 for about two to three dead times. It doesn't require the loop to be at steady state. However, if there is an upset that causes the ramp rate to change, the results will be inaccurate. In general, you should repeat this test in both directions and use the most conservative settings. Also, if the bump in controller output is much larger than the dead band, the shortcut method doesn't include the dead time from valve dead band. If the changes in controller output per scan approach the control valve dead band in size, you should add the additional dead time from the valve dead band to the observed dead time.

Figure 10 – Change in Ramp Rate for Non-Steady State

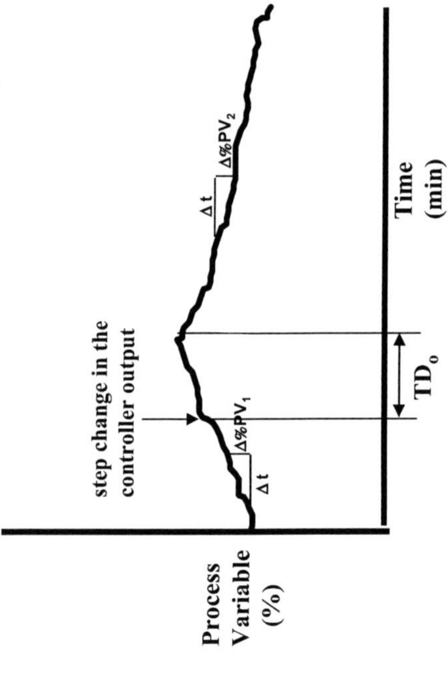

The shortcut method is also effective for pH loops because it can keep the test near the operating point on the titration curve. The use of a closed-loop method can get confusing for pH particularly if the oscillations develop into a limit cycle after being bounced back and forth between the flat ends of the titration curve. The period of such a limit cycle is extremely long and variable, and it will occur for a large range of controller gains.

A list of steps for the "Shortcut Method" are as follows:

1. Adjust the measurement filter to keep the controller output fluctuations caused by noise within the valve dead band.

2. Note the magnitude of output change for each reaction to typical upsets. With the controller in manual near set point, make a step change in the controller output ($\Delta\%CO$) of about the same magnitude as the output change you noted, but larger than twice the valve dead band.

3. Note the observed dead time and the change in ramp rates. If the process was lined out before the test, then the starting ramp rate is zero ($\Delta\%PV_1 / \Delta t = 0$).

4. Divide the change in ramp rate by the change in valve position to get the pseudo integrator gain (K_i). Then compute the dead time from the dead band.

5. Use the following equations. For a master or supervisory loop, omit τ_{dv}.

$$K_i = \frac{|(\Delta\%PV_2/\Delta t) - (\Delta\%PV_1/\Delta t)|}{|\Delta\%CO|}$$

$$K_c = \frac{K_x}{K_i \bullet \tau_{do}}$$

$$\tau_{dv} = \frac{DB}{K_x \bullet \Delta\%AVP} \bullet \tau_{do}$$

$$\Delta\%AVP = |\Delta\%CO| - \frac{DB}{2}$$

$$T_i = c_i \bullet (\tau_{dv} + \tau_{do})$$

$$T_d = c_d \bullet (\tau_{dv} + \tau_{do})$$

Where:

$\Delta\%AVP$ = change in actual valve position (%)
$\Delta\%CO$ = change in the controller output (%)
DB = dead band from valve hysteresis (%)
K_c = controller gain (dimensionless)
c_d = rate time coefficient (1.0 for back-mixed and 0.0 for plug flow volumes)
c_i = reset time coefficient (4.0 for back-mixed and 0.5 for plug flow volumes)
K_x = gain factor (1.0 for Ziegler-Nichols, 0.5 for IMC, and 0.25 for Lambda)
K_o = open-loop gain (dimensionless)
K_i = pseudo integrator open-loop gain (1/sec)
$\Delta\%PV$ = change in process variable (%)
Δt = change in time (sec)
τ = largest loop time constant (sec)
τ_{do} = dead time seen in open-loop test (sec)
τ_{dv} = dead time from control valve dead band (sec)
T_d = derivative (rate) time setting (seconds)
T_i = integral (reset) time setting (seconds/repeat)

Rule 15 – **Use the shortcut method when you want a quick estimate for a very slow or nonlinear loop, provided that the valve dead band is 0.25 percent or less and the step size keeps you near the operating point.** Make sure there are no load upsets during the test and that you measure the new rate of change of the PV for at least two dead times. You should repeat the test for both directions and use the most conservative tuning.

2.5 Simplified Lambda Tuning

If the loop is dead-time dominant, Lambda tuning is the best method. It also helps to minimize interactions and suppress oscillations. Lambda tuning is an open-loop method that is particularly effective for relatively fast loops, such as in pipelines, desuperheaters, static mixers, exchangers, conveyors, spin lines, and sheet (web) lines, or wherever there is plug flow. It provides a closed-loop time constant that approximates the open-loop time constant. Figure 11 shows the step change in controller output and the open-loop response of the process variable. The user simply needs to note these changes and the time required to reach 98 percent of the final response (T_{98}). This simplified form of the equation for Lambda

Figure 11 – Lambda Test

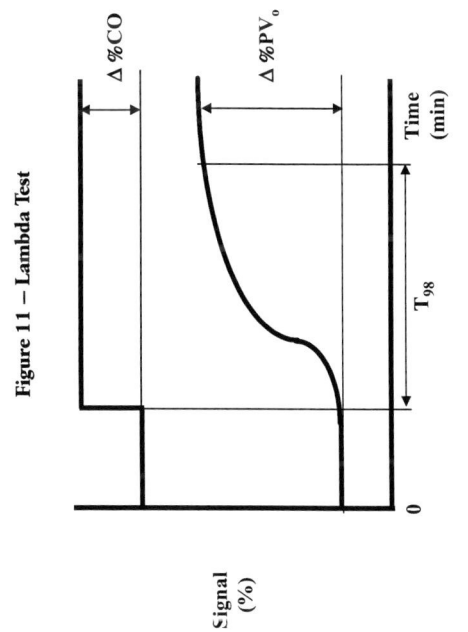

tuning is only suitable for self-regulating processes. Loop tuning software has expanded the actual Lambda tuning rules to cover a wide variety of processes.

The following is a list of steps for the "Simplified Lambda Tuning Method":

1. Adjust the measurement filter to keep the controller output fluctuations caused by noise within the valve dead band.

2. Note the magnitude of output change to determine the reaction to typical upsets. With the controller in manual near set point, make a step change in the controller output ($\Delta\%CO$) of about the same magnitude as the output change you noted, but larger than twice the valve dead band.

3. Note the observed dead time as the time it took to reach the first change in the process variable ($\Delta\%PV$) outside of the noise band.

4. Note the open-loop gain (the percentage change in measurement divided by the percentage change in controller output).

5. Note the response time (the time from the bump to 98 percent of the final value).

6. The reset setting (repeats/minute) is set equal to the inverse of one-fourth of the response time.

7. The controller gain is adjusted to be one-fourth of the inverse of the open-loop gain ($\lambda_f = 4$).

The equation for reset time setting is as follows:

$$T_i = T_{98}/4 = \tau_{do}/4 + \tau \text{ (seconds/repeat)}$$

The equation for gain setting is as follows:

$$K_c = (\Delta \%CO / \Delta \%PV) * (1/\lambda_f)$$

(λ_f is a tuning factor that is increased to provide a slower and smoother response.)

Rule 16 – **Use the Lambda tuning method for dead-time-dominant or fast loops that can be left in manual.** You should repeat the test for both directions and use the most conservative tuning.

Rule 17 – **In Lambda tuning, decrease λ_f to speed up the controller's response to upsets and set point changes (minimize Δ PV) and increase λ_f to slow down the response and reduce loop interaction (minimize Δ %CO).** This simplifies tuning to a single knob.

2.6 Set Point Response and Load Rejection Capability

One way to achieve a good set point response with a controller tuned for best load response is to add a set point lead-lag whose lag time tracks the integral time and whose lead time is adjusted to be a fraction of the lag time. Another alternative is to achieve a compromise by the use of the closed loop tuning method with about half of the gain that caused quarter amplitude cycling and an integral time (T_i) set as function of the quarter amplitude oscillation period (T_o) and the total loop dead time (τ_d) as follows:

$$T_i = T_o/[0.1 + 2 * \{(4 * \tau_d)/(0.7 * T_o)\}^2]$$

3.0 – Measurements and Valves

3.1 Watch Out for Bad Actors

The analog/digital converter (A/D) chatter from large temperature measurement spans and short scan times is the factor most frequently cited to explain why derivative action cannot be used in PID controllers. It is also a considerable source of dead time because the measurement must get out of the noise band if the PID controller is to discern a true load change from the chatter. There is a similar signal-to-noise ratio problem for level measurements, particularly if the signal is noisy due to sloshing or bubbles.

The next most frequent problem is sensor coating and drift. Resistance temperature detectors (RTDs) and smart transmitters can reduce drift by an order of magnitude. Keeping the velocity that is passing a sensor above 5 fps is the most effective way to keep a sensor clean.

Rule 18 – Narrow span ranges and slow scan times should be used for temperature and level measurements to minimize the

dead time and chatter from the analog/digital converter (A/D). For level measurements, you should minimize sensitivity to bubbles, sloshing, and coating.

Rule 19 – **Use RTDs and smart transmitters to reduce drift and calibration requirements.** They will more than pay for themselves through lower maintenance costs and more accurate process operating points.

3.2 Deadly Dead Band

According to pulp and paper plant studies, most loop variability is caused by poor tuning and control valve resolution. Figure 12 shows the dead band that occurs whenever a control valve changes direction and the staircasing that is the result of stick-slip action. The dead band and stick slip are usually largest near the valve seat. Valve resolution is the minimum change in signal in the same direction that will result in a flow change and is usually about half of the valve dead band. As you approach the resolution limit there is a dramatic increase in the loop dead time from dead band, as illustrated in Figure 13. This additional dead time can be about five times larger for pneumatic positioners than for digital positioners. Figure 14 shows that

Figure 12 – Dead Time from the Valve

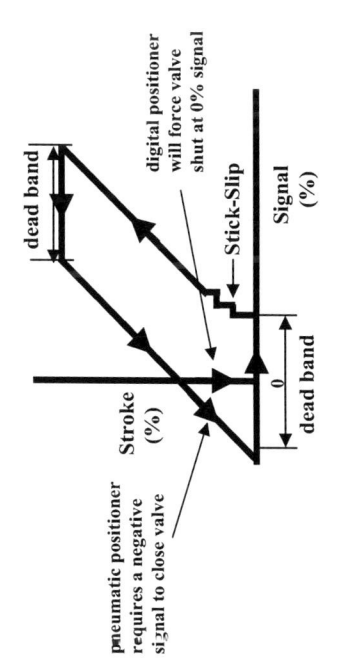

it is worse for piston actuators and rotary valves. Most valve manufacturers will choose a valve position and step size that is near the minimum shown in Figures 13 and 14. For large step sizes, the time required to move large amounts of air into or out of the actuator increases the response time.

Another problem with rotary valves is that the feedback to the positioner is from the actuator shaft. What might seem to be minor gaps in key-lock connections and twisting in long shafts actually means that the butterfly disk or ball may not move, even though the positioner sees a movement of the actuator shaft. To minimize this problem, you should use splined connections and short large-diameter shafts, and you should conduct actual flow tests on all rotary valves. All tests on control valves should use step sizes that approximate what is expected to occur as changes in the controller output from one scan to another ($< 0.5\%$). You should also use very large step sizes for pressure-relief valves and for compressor antisurge valves. There is nothing in a valve specification that requires that the valve will actually move. To ensure that a valve will respond to meet the needs of a control loop, you should add the dynamic classes in Table 4 to the valve specification sheet when purchasing a control valve.

Figure 13 – Dramatic Rise in Response Time as You Approach the Resolution Limit

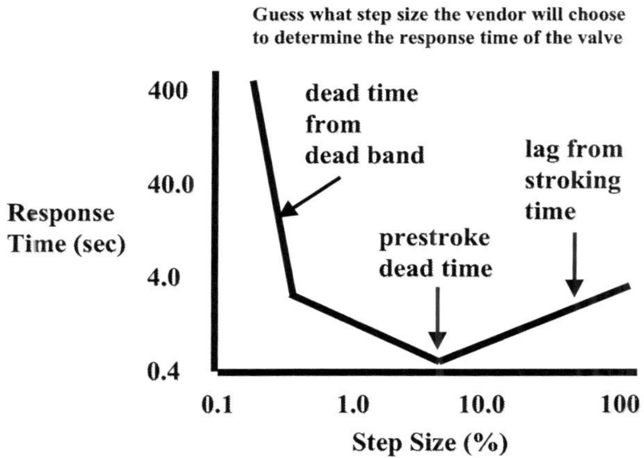

Guess what step size the vendor will choose to determine the response time of the valve

MEASUREMENTS AND VALVES

Figure 14 – Response Time Depends upon the Valve, Shaft, Size and Connections, Actuator, and Positioner

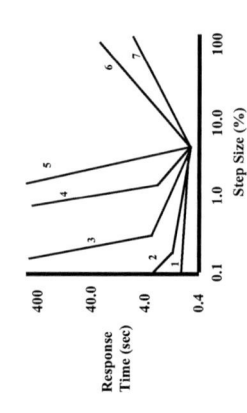

1 - sliding stem valve with diaphragm actuator and digital positioner with pulse width modulated solenoids.
2 - sliding stem valve with diaphragm actuator and digital positioner with nozzle flapper
3 - sliding stem valve with diaphragm actuator and pneumatic positioner
4 - rotary valve with piston actuator and digital positioner
5 - rotary valve (tight shutoff) with piston actuator and pneumatic positioner
6 - very large rotary valve (>6") with any type of positioner
7 - sliding stem valve with a digital positioner

Table 4 – Dynamic Classes of Control Valves (Four classes A → D for each of the four categories 1 → 4)

1-Minimum Step Classes	2-Maximum Step Classes
Class A - 3.0% ± 0.3%	Class A - 5% ± 0.5%
Class B - 1.0% ± 0.1%	Class B - 10% ± 1.0%
Class C - 0.5% ± 0.1%	Class C - 20% ± 2.0%
Class D - 0.2% ± 0.1%	Class D - 50% ± 2.0%
3-Response Time Classes	**4-Minimum Position Classes**
Class A - 15 sec	Class A - 30%
Class B - 5 sec	Class B - 20%
Class C - 2 sec	Class C - 10%
Class D - 1 sec	Class D - 0%

Note: The response time is the time it takes the trim (not the actuator) to stay within 10 percent of step or within 0.1 percent of span, whichever is largest. (Overshoot is OK if recovery is within this offset from the desired position of the trim)

Referring to Table 4, consider a DACB class control valve. It will respond to signals larger than 0.2 percent and smaller than 5.0 percent in less than 2 seconds above a position of 20 percent. If you don't care about how well a valve responds, specify class AAAA--almost any valve will meet it.

A general-purpose application would be BAAB. Examples of other classes for various types of control loops are given in the following list. To choose a typical class, use the loop type that actually throttles the control valve. For cascade loops, you should use the slave loop.

- Vessel or Column Temperature Control => CABC
- Exchanger Temperature Control => CACC
- Pipeline Temperature Control => CCCC
- Liquid Pressure or Flow Control => CACC
- Vessel Level Control => CABC
- Gas or Steam Pressure Control => CBDC
- Compressor Surge Control => CCCD
- Vessel pH Control => DABD
- Pipeline pH Control => DBCD
- Pressure Relief => BCCD
- For split range, use class D for last category (minimum throttle position)

The best way to see if the flow actually changed for a small change in controller output is to make a sensitive flow measurement with very low noise. This is particularly important for rotary valves or pneumatic positioners. Another clue that the valve has a problem is if the control loop oscillations are smaller in manual than in auto (this assumes you have tuned the controller). If it is a critical rotary valve and there is no flow measurement, you need to take the valve out of the pipeline. In the shop, you need to then attach a travel indicator to the butterfly disk or ball to check whether they track the changes in the actuator shaft.

Finally, the change in the flow measurement for a change in controller output will also show if the control valve is on a portion of the installed characteristic that is too steep or flat. If there is no flow measurement, ask the valve manufacturer to generate an installed valve characteristic for your piping system. Signal characterization to linearize the valve will decrease valve dead time from its resolution limit for the flat portion of the curve, but it will increase this dead time for the steep portion of the curve.

Figure 15 summarizes the maintenance tests for a good or bad valve. The change in PV can be any process variable, but it works best if it is a fast measurement such as flow or pressure. Slow process variables such as temperature and level respond much too slowly, especially for small changes in controller output, to make it practical to determine valve dead band.

Rule 20 – **To achieve the tightest control (i.e., least variability), use sliding stem valves and digital positioners.** If you use rotary valves, you should have a flow

Figure 15 – Maintenance Tests for a Good or Poor Control Valve

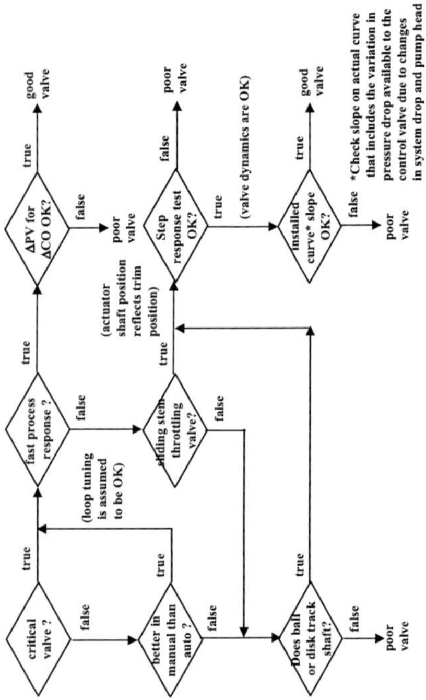

*Check slope on actual curve that includes the variation in pressure drop available to the control valve due to changes in system drop and pump head

measurement, splined shaft connections, and a short large-diameter shaft.

Rule 21 – **If you must operate on the flat portion of an installed valve characteristic, use signal characterization to make the valve more sensitive.** Be careful to avoid increasing the valve dead time from dead band on the steep portion of the curve. Also make sure you don't eliminate an equal-percentage characteristic used to compensate for a flow ratio gain.

In the normal scheme of things, slip is worse than stick, stick is worse than dead band, and dead band is worse than stroking time. For sliding stem valves, stick-slip goes hand in hand with dead band since the common cause is excessive packing friction. Rotary valves can have a large dead band from gaps in connections and linkages and not much stick-slip. However, high performance valves generally have significant stick-slip and shaft windup from excessive packing and seating friction. Stick-slip causes a limit cycle in all loops, dead band causes a limit cycle in integrating loops, and a large stroking time causes instability when the rate of change of the controller output from integral action is faster than the slewing rate of the valve.

3.3 Sticky Situations

For those of you who have been stuck with dead band and stick-slip and want to break free, this section will discuss the sources of the problems and details of the solutions.

Tight connections are the goal for the transmission of force from the actuator to the internal ball, disk, or plug. For a sliding stem valve there is a single connection between the actuator shaft and the stem of the plug. In a rotary valve there is a connection between the actuator shaft, valve shaft, and ball or disk stem. There may also be linkages associated with the translation of vertical actuator motion to rotary valve motion. The play in each connection, which is usually a pinned, keyed, or splined slot, determines how much dead band (backlash) exists whenever the valve needs to reverse direction. While the play of loose slots may be fun at Vegas, they are deadly in control loops.

Manufacturers have been doing a better job of not only improving the tightness of connections but also in increasing the number of pins, keys, and splines around the circumference of the shaft to increase the effectiveness of the transmission of force. A splined shaft with a large number of tight spline connections has the least backlash.

In rotary valves, the positioner feedback measurement is often the actuator shaft. The tightness of the intervening connections determines whether the positioner is actually seeing what the ball or disc is doing. On top of this, there may be a scotch yoke actuator where the looseness of the gear mesh adds another degree of slop in valve response.

> *Rule 22* – To reduce the extra dead band in rotary valves, use splined shafts and digital positioners that measure the actual rotation of the internal ball or disk.

For surfaces that must slide, the tightness and roughness of these surfaces leads to high friction, commonly called *sticktion*. The smallest change in signal that a control valve can respond to regardless of its direction or past movement is called *stick*. When the valve does move, it slips where the incremental change is approximately equal to the stick. Excessive break away torque, inadequately sized actuators, operation near the seat, and poorly tuned positioners can result in a slip much larger than the stick even though the packing friction is not too high. Control loops will oscillate regardless of tuning with an amplitude equal to the slip multiplied by the open loop gain. The resolution capability or sensitivity of the valve is set by the slip, even though it is commonly defined based on the stick.

Here is the rub: Until valves can be made without sliding surfaces, there will be friction and a stick and slip that are at least as large as the manufacturer's stated resolution or sensitivity limit. For the best designed valves in perfect condition, the resolution is about 0.05%, which approaches the resolution limit of a 12-bit analog/digital (A/D) converter with a sign bit. However, most valves are far from perfect and the environment in process applications is not any where as nice as in the valve shop or test lab. Consequently, users have lowered their expectations and consider 0.5% resolution to be a good valve. The following situations can make reality reek by increasing stick-slip and achieving even a 0.5% resolution seem like nirvana.

1. Crud on the stem, trim, or seals that reduce the slickness of surfaces

2. High temperatures where the expansion reduces the clearances between surfaces

3. Shaft windup in rotary valves where the shaft twists and then springs free (jumps)

4. Plunges in rotary torque requirement upon break away when the disc jumps open

5. Tightened or roughened packing surfaces

6. Low performance or poorly tuned positioners

An isolation, safety interlock, or batch sequencing valve designed for tight shutoff should never be used for a control valve and vice versa. On-off valves do not make good throttling valves even though a positioner has been added. On-off valves are designed to be either completely open or closed. Control valves are designed to move freely near the closed position. Control valve specification sheets that have a leakage class but not a stick-slip requirement, leads to the selection of on-off valves.

While it is best to reduce the sources of high friction, the stick-slip can be reduced by installing an actuator with plenty of muscle and a digital positioner tuned with a high gain and rate setting.

Rule 23 – **To reduce stick-slip, avoid valves designed for tight shutoff and use low friction properly tightened packing, gener-**

ously sized actuators, and aggressively tuned digital positioners.

3.4 Fouled Sensors

A coating on a thermowell will generally reduce its heat transfer coefficient and provide a thermal insulating barrier to the fluid temperature. The response of the sensor slows down and the temperature difference between the inside tip and process fluid increases, which translates to a bigger lag time and dynamic error in the temperature measurement. For a ramping temperature change, the measured temperature will be delayed by this lag time. Also, if the lag time is more than 25% of the total dead time in the loop, a controller tuned for performance rather than robustness may go unstable. The controller gain must be decreased and the reset time increased to deal with the additional sensor lag time.

Since temperature control systems tend to have too small of a reset time to begin with per the Figure 2 heating and cooling example, the most noticeable symptom for a fouled sensor is a reset cycle and a longer period. However, the sensor lag acts like a signal filter or a transmitter dampening setting and attenuates noise and disturbances seen by the loop. Thus, the amplitude of noise (best checked in manual) and of oscillations for upsets will be smaller for a fouled sensor. The appearance of a flatter trend has fooled many individuals. A biochemist once showed me how he smoothed out the trend chart of a fermentor temperature by partial withdrawal of a temperature sensor in its thermowell. The resulting air gap acted as an excellent thermal barrier. Spring loaded temperature sensors are used to ensure the tip of the RTD or thermocouple touches the bottom of the thermowell.

Rule 24 – **If the oscillations in a loop become more frequent and slower for the same tuning, check to make sure the sensor response has not slowed down from a coating.**

A pH electrode is incredibly sensitive to coatings. A 10 millimeter thick film can increase the lag time from 5 seconds to 100 seconds. The coating acts as a barrier to the migration of hydrogen ions that need to get the surface of the glass electrode to change its millivolt potential. The coating can be so thick that the electrode does not respond at all.

The time to about 63% of the final response for the insertion of a temperature sensor in a bath or an electrode in a buffer is the sensor lag time. Washing or decontamination of the sensor and bath or solution itself may remove most of the coating. Thus, removing the sensor for testing in a shop often does not disclose the severity or nature of the coating. If safety procedures allow it, the sensor surface should be inspected as it is removed.

> *Rule 25* – **If safe and practical, visually inspect sensor surfaces for coatings as they are removed and before they are washed and sent to the maintenance shop.**

Coatings are less likely to occur if a high fluid velocity sweeps parallel to the surface. The threshold velocity that is reported to make a discernable difference in the rate of coating is about 5 feet per second, which is typical for a pipeline at normal flow. The velocity in the most highly agitated reactors is rarely more than 1 foot per second. Higher velocities also reduce the lag time of clean thermowells and electrodes. Thus thermowells and electrodes installed in pipelines rather than vessels stay cleaner and have a faster response clean or dirty. However, if the flow is throttled or stopped, coatings will develop that may not be able to be cleaned by a return to normal flow. Sticky coatings are difficult to remove and increase the susceptibility to further coating. The key is to never let the coating start. The installation of thermowells and electrodes in recirculation lines is the best insurance because these lines tend to have a constant high flow rate. If the fluid is abrasive, the velocity may need to be lowered by the addition of a wide spot in the line.

Rule 26 – **Install thermowells and electrodes in pipelines rather than vessels to reduce fouling.** If the process flow stops, drain and flush the line to avoid having the sensor sit in a stagnant process fluid. Recirculation lines are more likely to keep the velocity high and constant at the sensor.

Flat surface electrodes get the same cleaning action for a much lower velocity. The primary flow pattern should be parallel to the surface for the best sweeping action.

3.5 Noisy Measurements

If the sensor location is too close to the discharge of an exchanger, desuperheater, pump, or static mixer, the response may be noisy because streams have not had the time to mix or bubbles or solids time to dissolve. On the other hand, the distance between the equipment and the sensor adds a transportation delay equal to this volume divided by the flow. Therefore, the optimum location is about 5 to 10 pipe diameters downstream of the equipment. Strong acids and bases require twice the distance because of the extreme process sensitivity (gain). Bubbles and particles may require even longer distances. The bubble size decreases and its ability to dissolve increases as the velocity increases.

Sometimes overlooked is making sure the tip of the sensor is near the middle of the pipeline to see a more representative and smoother temperature or concentration. This is particularly important for pipes with bubbles, particles, high viscosity, or jackets. Installing a thermowell in an elbow keeps the insertion length in the center of the pipe. Since electrodes don't have conduction errors like thermowells, maximizing the insertion length in the process is not important. However, pH electrodes require that an electrode be installed above horizontal so that the internal air bubble does not get lodged in the tip. Most manufacturers designate a 30 to 45 degree angle from horizontal as optimal.

Rule 27 – **Install thermowells and sensors within 5 to 10 pipe diameters of the equipment discharge with the tip near the center of the pipeline to minimize noise.**

For a thermowell, the best location is an elbow. A pH electrode must be angled above horizontal. Bubbles and strong acids and strong bases require longer distances.

Differential head, thermal mass, and vortex flowmeters will have noisy signals unless the flow profile is uniform. The number of pipe diameters of straight run needed upstream can vary from 10 to 40 pipe diameters depending on the location of valves, fittings, and elbows upstream. Flow straightening devices can significantly reduce the straight run requirements. A control valve should never be located upstream of a flowmeter. The straight run needed downstream is generally 5 pipe diameters. The straight run specifications for magnetic flowmeters are just a couple of diameters. Coriolis flowmeters have no official requirement, but it is not a good practice to bolt them to control valves or equipment.

Rule 28 – **Ensure differential head, thermal mass, and vortex flowmeters have enough straight pipe runs upstream to see a good**

flow profile that reduces noise. Always install the control valve downstream of the flowmeter.

Operating near the low limit of the allowable range for the sensor can cause a noisy response. This is particularly a problem for differential head flowmeters. Changes in Reynolds number can change the meter coefficient, and the transition zone between turbulent and laminar flow can cause erratic behavior. If you go below the limit for vortex or magnetic flowmeter, the measurement signals can go bonkers. The rangeability of vortex meters and magmeters depends on the fluid velocity in the meter and the kinematic density and conductivity, respectively.

Rule 29 – Make sure the meter size and range for the fluid conditions ensures the operating range is always above the low rangeability limit of a differential head, magnetic, thermal mass, and vortex flowmeter.

4.0–Control Considerations

4.1 Auto Tuners

The relay method auto tuner shown in Figure 16 is effective and simple. It keeps the loop under control by constantly forcing the PV back and forth across the starting point of the test. Its effectiveness depends upon the noise band being set correctly so that fluctuations from noise are not measured as loop oscillations. Figure 17 shows a single step in controller output that uses the shortcut tuning method to provide an estimate of tuning settings and then a series of successive steps to provide an automatic identification of valve dead band.

4.2 Uncommonly Good Practices for Common Loops

For flow and pressure loops, first make sure the scan time is fast enough. For gas pressure, the gain must be maximized. For level loops it is most important that you decide whether the level control must be tight or loose and that you minimize the reset action. For temperature loops it is critical that you use a narrow calibration span and slow scan time. For concentration loops you must make sure that the sensor is not

Figure 16 – Relay Method Auto Tuner

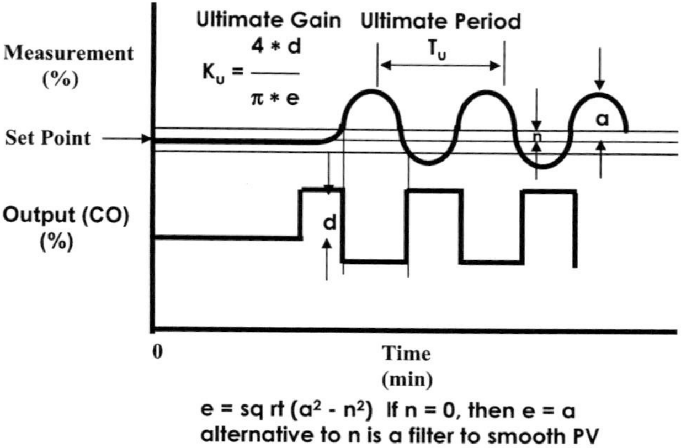

Measurement (%)
Set Point
Output (CO) (%)
Ultimate Gain
$K_U = \frac{4 * d}{\pi * e}$
Ultimate Period
T_U
a
n
d
0
Time (min)
e = sq rt (a² - n²) If n = 0, then e = a
alternative to n is a filter to smooth PV

Figure 17 – Enhanced Pretest and Pretune

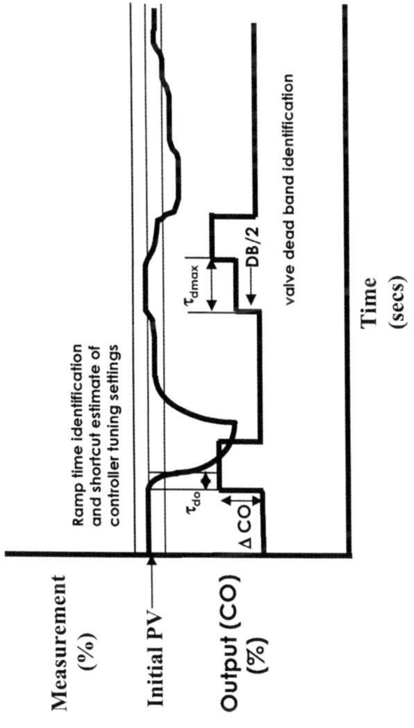

coated (velocity > 5 fps) and that the signal is fast (minimum transportation delay) and smooth (no noise). For pH loops with an operating point on a steep titration curve, the greatest need is for an exceptionally precise control valve (resolution < 0.15%). Characterizing the signal of the process variable according to the titration curve can help reduce the oscillations.

4.3 Dead-Time Compensation and Warp Drive

When I first left home, my dad said, "Be as honest as the day is long, don't talk when you should listen, and don't be fooled into thinking a dead-time compensator can eliminate process dead time." A dead-time compensator such as a Smith Predictor can cancel out the effect of dead time for changes in the controller output and make possible higher gains and faster reset action (smaller reset times). However, it is a common misconception that it eliminates dead time when correcting disturbances from feedback action. The minimum peak error still corresponds to the excursion of the process variable in one total loop dead time. The disturbance and correction must still make a complete traversal of the block diagram. Dead time in the plant cannot be eliminated by an algorithm without violating the laws of physics. Unless you have Scotty and warp drive, you are stuck with the dead time caused by equipment, piping, instrumentation, and control valves in your loop.

A dead-time compensator is very sensitive to an overestimate of the dead time. A dead time that is 25 percent larger than actual can cause instability. A major source of dead time is a transportation delay, which can be computed for a pipeline as the volume divided by the flow or for a sheet as the distance divided by the speed. Overestimates of the process gain and underestimates of the time constant are also problems but to a lesser degree.

To derive the full advantage of a Smith Predictor, you should increase the controller gain and the reset action. For a negligible time constant, the reset time can be set equal to one-fourth of the uncorrected dead time to provide a reset action that is an order of magnitude greater. A dead-time compensator is most effective on dead-time-dominant loops where the dead time can be accurately calculated and updated in the predictor.

Most dead-time compensators can be reduced to some form of the Smith Predictor shown in Figure 18. The PID controller output passes through a dead time, single time constant, and steady-state gain model of the valve, process, and sensor dynamics. The model output with and without the loop dead time is subtracted from and added to the measurement of the process variable, respectively. This leaves the model output without the dead time as the controlled variable. The dead time has been removed from the loop as far as changes in the controller output are concerned. Note that the controlled variable is no longer the actual process variable.

Rule 30 – **To maximize the stability of a dead-time compensator, update the dead time as quickly and accurately as possible and make sure you always underestimate the dead time and process gain and overestimate the time constant.** You should include the change in dead time resulting from a change in a transport delay by using a computation in which the dead time is inversely proportional to flow or speed.

Figure 18 – Smith Predictor

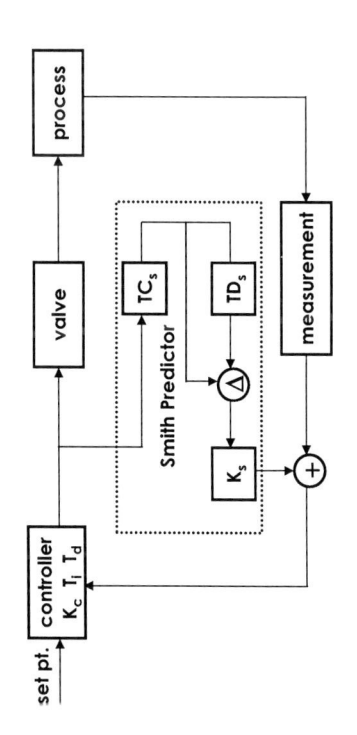

Rule 31 – **The reset time should be set to be about one-fourth (¼) of the uncompensated dead time to get the most out of a dead-time compensator.** The reset time can be decreased by an order of magnitude by adding a dead-time compensator with a 10 percent or less underestimate of the dead time.

4.4 I Have So Much Feedforward, I Eat before I Am Hungry

A feedforward signal that is accurate both in gain and in timing can make an impressive impact on control loop performance, especially for unit operations with large dead times such as distillation columns and large reactors. Ideally, the compensating effect from the feedforward signal should arrive in the process at the same time as the load change and be equal but opposite to the upset. For fast loops, the timing is tight, and the feedforward signal may arrive early and cause an inverse response. If it is too late, the feedforward signal creates a second disturbance. For fast loops, the timing is more critical than the gain, but dynamic compensation is often neglected. Figure 19 shows the patterns of response that are a clue to how you should adjust the feedforward gain and timing.

Figure 19 – Feedforward Tuning

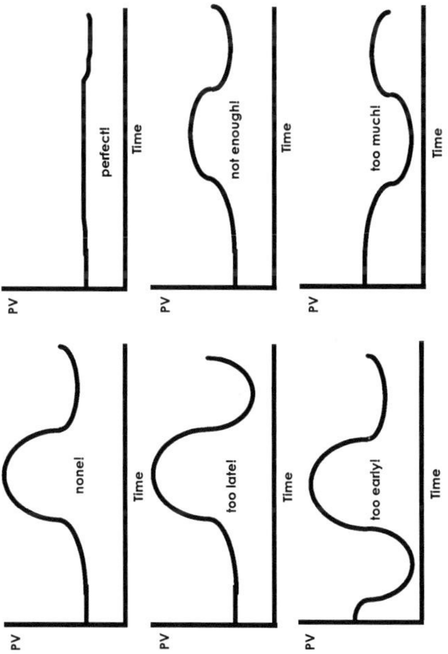

CONTROL CONSIDERATIONS

You can greatly improve a loop's set point response by adding the percentage change in set point into the controller output as a feedforward signal that has a gain of about half of the controller gain, an action opposite to the control action, and a slight filter. This provides the kick you would normally get from a high gain setting. The loop can then rely on reset action to make the rest of the transition to the new set point.

Rule 32 – **Both the feedforward gain and the dynamic compensation (dead time and lead lag) must be set properly to get the most benefit.** For fast processes the timing is more critical. It is better for the signal to arrive a little late than too early because inverse response is extremely disruptive.

Rule 33 – **Add a feedforward signal of a set point in percent to the controller output with a gain about half of the controller gain, an action opposite of the controller action, and a small filter.** The feedforward action is opposite the controller action because the controller action works on error, which is the measurement minus the set point. For nonlinear valves, you may need to add signal characterization.

4.5 Cascade Control Tuning

Cascade control is a type of control in which a secondary (slave or inner) loop is added that gets a set point from a primary (master or outer) loop. If the secondary loop response (both dead time and time constant) is five times faster than the primary loop response, there is no interaction between the loops, and the secondary loop can correct for upsets it can measure before they affect the primary loop. If the secondary loop is not fast enough, you must increase the scan time and the PV filter time of the primary loop or you will have to decrease the controller gain and rate action of the primary loop to reduce the interaction. Besides catching an upset quicker, the secondary loop may also help linearize the response of the primary loop. For example, in the cascade of reactor temperature to coolant jacket exit temperature, the secondary loop of jacket temperature will make the process gain for reactor temperature linear in addition to sensing coolant temperature disturbances before they affect the reactor temperature. Similarly, the cascade of exit coolant temperature to a flow loop on coolant makeup will remove the nonlinearity of the installed characteristic and dead band of the valve in addition to reducing the effect of coolant pressure changes. The most common type of secondary loop is the flow loop. It can considerably improve the performance of a primary loop for concentration, level, and temperature control. The flow loop is not generally recommended for liquid pressure control because the speed of response of the liquid pressure and flow are about the same. Another type of cascade control that should be used more frequently is the cascade control of still, reactor, or evaporator temperature to steam pressure control. In this kind of control, the steam pressure loop compensates not only for steam supply pressure upsets but also for changes in the condensing rates (heat load and transfer) as reflected in the steam coil or jacket pressure.

You should tune the secondary loop for an immediate response before tuning the primary loop. An offset in the secondary loop is usually of no consequence since the sole purpose of the secondary loop is to meet the demands of a primary loop. You should tune the secondary loop with maximum gain action and minimum reset action and use a feedforward of the set point. This is particularly important for cascade control of gas pressure control to flow. If the abrupt changes in the secondary loop output upset another important loop, then you may need to use Lambda tuning to make the secondary loop response smoother and more gradual.

The most common mistake is to forget to properly set the output limits of the primary loop. You must set the output limits on the primary loop so they match the set point limits on the secondary loop. Also, the primary loop must not wind up when the secondary loop output is at its output limits.

Rule 34 – **If the dead time and time constant of the primary loop are not five times slower than they are for the secondary loop, the primary loop must be slowed down.** The scan time and filter time must be increased or the controller gain and rate action must be decreased in the primary loop.

Rule 35 – **You must tune the secondary loop first and use gain and/or set point feedforward action in it so the secondary loop can immediately respond to the set point changes from the primary loop.** Reset action is too slow to be the major source of control action in the secondary loop. If you use a feedforward of the set point, make sure it is scaled properly. If the secondary loop fights with other important loops, use Lambda tuning to minimize the interaction.

Rule 36 – **Make sure the output limits of the primary loop match the set point limits of the secondary loop.** In a fieldbus-based system, the primary loop output limits use the engineering units of the secondary loop. In older systems, the primary loop output limits are usually expressed as a percentage of the secondary loop scale.

4.6 Keep the Secondary Loop on the Move

A common mistake is to tune a secondary loop for a smooth set point response in a cascade loop with a large process time constant. The load rejection capability of the secondary loop is much more important. In particular, it is the rise time rather than the settling time of the secondary loop that should be minimized to make the initial response as fast as possible. The remote set point (RSP) of a secondary controller is often changed by the primary controller before the secondary loop PV even crosses its RSP. The use of a filter on the RSP of a secondary loop will seriously degrade cascade control.

5.0 – Troubleshooting

5.1 Patience, Heck, I Need to Solve the Problem

Figures 20 and 21 show some diagnostics for loops in manual and auto. In these diagrams, reset action (repeats per minute) is referenced rather than reset time (seconds per repeat). If there are fewer oscillations with the loop in manual, the problem is either a poor control valve or controller tuning. If there are also fewer oscillations in other loops, then they are probably caused by an interaction between this loop and other loops. If the oscillations persist and are fast, they are probably due to electromagnetic interference (EMI), sensor noise, pressure waves, or resonance. If the oscillations persist and are slow, then they are periodic upsets from other loops that have poor tuning or valves or are caused by on-off actions (level switches), steam traps, pressure regulators, burps (column flooding), or flashing. If the oscillations dissipate when the valve is closed, the oscillations were caused by pressure fluctuations at the valve.

With a loop in manual, you can find the valve dead band by making a step change of 0.25 percent in the valve and then waiting more than the dead time to see if the actual process variable responds to the change in the controller output. For slow processes or in situations where you are reasonably sure the valve is really lousy, you can increase the step size to 0.5 percent or even 1.0 percent. This will allow the test to be completed before your patience wears thin. If there is no response, repeat the step. The steps should all be in the same direction until there is response of the process variable that is outside of the noise band. Then repeat test for the opposite direction. The absolute magnitude of the total number of step changes needed to get a response in both directions is a measure of the valve dead band.

Rule 37 – **If the oscillation goes away when you put the loop in manual, then the loop is the cause of the oscillation.** The culprit could be the control valve, tuning, or loop interaction.

Rule 38 – **If the oscillation only goes away when you close the valve, then it was caused by pressure fluctuations at the valve.** The culprit could be an oscillating pressure loop, on-off actions, or oscillating users on the same header.

> *Rule 39* – **To track down the source of an oscillation, put each loop in manual and stop each on-off action one at a time.** When the oscillation stops, you know the culprit was the last loop you put in manual or the last on-off action you stopped.

If the loop is in automatic and there are fast periodic upsets, the oscillations are probably caused by EMI, sensor noise, pressure waves, or resonance. If you experience slow periodic upsets and a period much greater than four times the dead time, then you should suspect on-off actions, steam traps, pressure regulators, burping, or flashing as the source. Otherwise, tuning is probably the culprit. When PV wanders in automatic it is a symptom of the control loop operating on the flat portion of an installed valve characteristic. If the PV falters for a dead-time-dominant loop (see Figure 3b), it is an indication of too much gain action. If the PV staircases (see Figure 5), it is an indication of too much rate action. If the PV overshoots and develops a slow oscillation whose period varies with the reset setting (see Figure 4), it is a sign of too much reset action (i.e., the reset time too small). If the loop period is very sensitive to reset action, it is a sign that there is excessive valve dead band. For sticky valves, you can make the oscillation significantly faster by using more rate and gain action. This will get the controller output through the dead band more quickly.

If the period of an upset is near the natural period of the loop, which is two to four times the dead time, then the loop will amplify the disturbance, and the oscillation will be worse when the loop is in automatic. If the period is much less than the natural period it is uncontrollable noise. For a loop to have a chance of attenuating the upset, the period of the upset must be more than twice as large as the natural period of the loop.

Note that on Figures 20 and 21 "reset too hi" means the reset action is too high, which means the reset time is too small.

Rule 40 – **If the oscillation period for the loop in automatic drastically increases for less aggressive controller tuning, the valve has excessive stick-slip.** It takes longer for the controller output to work through the resolution if the PID action is slower due to less gain, reset, and rate action.

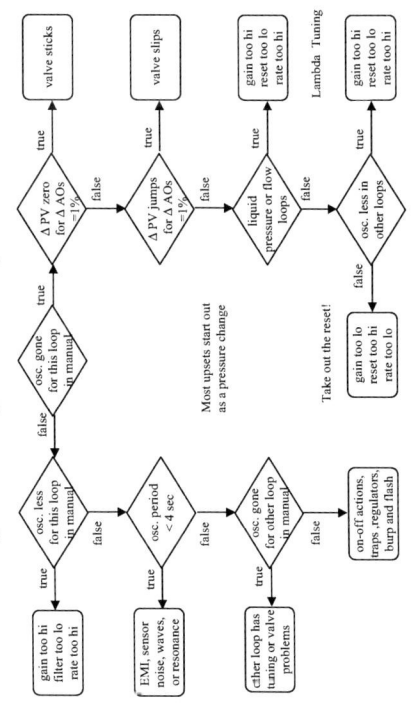

Figure 20 – Diagnostics for Loops in Manual

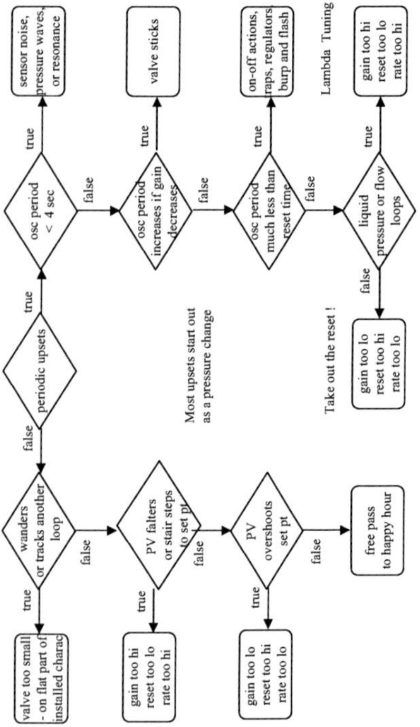

Figure 21 – Diagnostics for Loops in Automatic

The oscillation can be made much faster and tighter by using higher gain and rate action.

Figure 22 is a spectrum analysis for a level, which shows that the predominant frequency is 0.08 cycles per minute. By analyzing the power spectrum of process variables in the application depicted in the figure, Walsh Automation was able to find the source of a cycle with a twelve-minute period in a series of distillation columns. It was an oscillation in the feed, which was caused by level switches on a feed tank that turned pumps on and off. This power spectrum analysis tool can quickly point to common periodic oscillations that noise and upsets would make it difficult to spot in trend recordings.

Rule 41 – **A power spectrum analyzer can rapidly point you to the culprit by indicating which loops experience significant peaks in the power at the same frequency.**

The first step is to enter the data for loops in automatic into the power spectrum analyzer. The data gathering must be done quickly enough to prevent aliasing. For chemical processes, it is sufficient to use data from a historian with no compression and an update time of 1 second. For sheets (webs), you may need to store the data using a device with an exceptionally fast

Figure 22 – Power Spectrum Analysis (Source: Courtesy of Walsh Automation, Toronto)

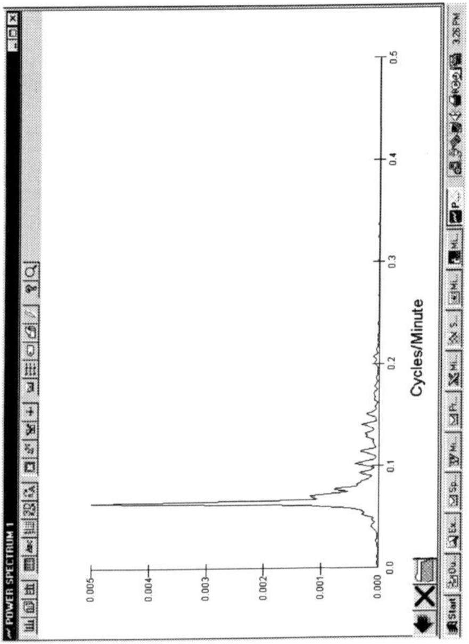

scan time (50 milliseconds or less) that is directly connected to the controlled variable terminals.

5.2 Great Expectations and Practical Limitations

When all is said and done, a loop will at best only pass on variability from its PV to its output. Variability in the output of one loop can greatly disturb another loop. Loops will actually increase the variability of the PV if there is a sticky valve, overly aggressive tuning settings, or upsets that have a period close to the natural period of the loop (two to four times the loop dead time). Most periodic upsets are caused by on-off actions and oscillating loops (which are often associated with level switches or level controllers that have too much gain or reset action). You can avoid a lot of headaches and wasted effort by first resolving valve problems and periodic upsets.

The most troublesome practice is using reset action correctly. It is not widely understood that the reset time as a factor of the loop period varies by an order of magnitude depending upon the degree of dead-time dominance and self-regulation. For example, for a process that tends to ramp or run away, the reset time should be larger than ten times the loop period. For a self-regulating process with a small dead-time-to-time-constant ratio, the reset time should be between 0.5 and 1.0 times the loop period. For pure dead-time processes, the reset time can be one quarter (1/4) of the dead time. Since the natural period is twice the dead time for a dead-time-dominant loop, the reset time ends up being about one-eighth (1/8) of the loop period. If a dead-time compensator is added to this loop, the reset time can be set to be about one quarter of the uncompensated dead time. If the dead time is underestimated by 10 percent, the reset time can be as low as one eightieth (1/80) of the loop period for a dead-time compensator.

To make it even more interesting, the rule of thumb to set the reset time greater than ten times the loop period for integrating and runaway processes is designed to essentially eliminate the possibility of oscillations due to reset action. There is a gain window for such non self-regulating processes. If the gain is below the low limit or above the high limit, there is a loss of control. The closer the controller is to the low gain limit, the more sensitive it is to reset action. Proportional action makes the response more self-regulating and thus, more tolerant of reset action. Some consultants recommend proportional-only controllers for level loops and proportional plus derivative controllers for exothermic reactor temperature loops to protect against instability caused by reset. In practice, some reset is useful to eliminate offset and facilitate startup. For level loops, it has been found that to prevent oscillations, the product of the controller gain and reset time must be greater than 4 times the fastest time for the level to ramp full scale. While the actual minimum product of gain and reset that triggers oscillations will change depending upon the amount of loop dead time and valve dead band, the relationship is useful to determine how to simultaneously adjust the gain and reset for level loops. If the loop response is smooth and the controller gain is doubled, the reset time can be halved and still keep the product of the two adjustments the same. It is critical to note that the opposite is true (the reset time must be doubled) if the loop is approaching the upper gain limit. If you are confused, don't feel alone. It is safe to say that 99% of users who don't carry this guide don't even have a clue that this strange twist of the rule on tuning reset even exists.

The following equation by Harold Wade shows how to eliminate oscillations near the low gain limit for a level controller if the dead time and valve dead band are negligible:

$$K_C * T_I > 4 * T_L$$

$$T_L = V_S / \Delta F$$

Where:

ΔF = maximum difference between inlet and outlet flows (gpm)
K_C = controller gain
T_I = controller reset time (minutes/repeat)
T_L = fastest level full scale ramp time (minutes)
V_S = volume for level measurement span (gals)

Finally, in order to tune the loop you need to be able to see the response. If the update time is too slow, the period of an oscillation and any time measurements will appear longer than actual. If 10% accuracy is desired, then the update time must be less than one tenth of the dead time or time constant. The scan times of any devices that supply data should be twice as fast as the trend update time to ensure sufficient over sampling. In order to get good accurate estimates of gains, the compression (value for exception reporting) should be smaller than the measurement or valve resolution. Otherwise the gains will be in error to the degree that the compression approaches the step size.

Rule 42 – **The update time of trend recordings should be less than one tenth of the loop dead time or time constant.** The compression of data should be less than one tenth of the step size and less than the measurement or valve resolution, whichever is largest.

6.0 – Tuning Requirements for Various Applications

This section hits the highlights of what are the major considerations in tuning loops in actual applications. The biggest most common problem is that the more critical control loops in the process industry are slow (long dead times and large time constants). The tuning tests for these loops take so long they are vulnerable to unmeasured load upsets during the test. Particularly problematic are large distillation columns where the time to steady state exceeds a shift and batch operations where there is no steady state or the time to steady state approaches or exceeds the batch cycle time. The shortcut tuning method that looks only at the initial change in ramp rates and does not need to wait for a steady state is the most reliable manual procedure for these processes.

Always keep in mind the following six points when approaching every application:

1. The step size in output or set point for a tuning test should be at least five times the noise band, valve dead band, and unmeasured load upsets.

2. Unmeasured disturbances will deform the response and stick-slip will cause a sustained saw tooth oscillation in the controller output of a self-regulating loop and in the controlled variable of an integrating loop.

3. When rounded oscillations appear in a loop, they are usually the result of poor tuning or interaction of loops. To find the culprit loop, put controllers one at a time in manual until the oscillations stop,

4. Processes are nonlinear. The process gain, dead time, and time constant will change with operating point and load so any attempt to get more than one or two significant digits in tuning settings is an exercise in frustration.

5. Most loops where the process time constant is larger than the dead time have too much reset action (too small of a reset time).

6. Most loops where the dead time is larger than the process time constant have too little reset action (too large of a reset time).

While manual tuning methods are noted in this section, an auto tuner can be used and the tuning rule associated with the suggested method selected. For example, a relay tuner or adaptive controller can be used and either Lambda or typical PID tuning rules chosen. For PID tuning, the normal Zeigler-Nichols factors applied to the ultimate period and ultimate gain are adjusted per the general-purpose closed loop method. Specifically, the controller gain is approximately cut in half to provide more robustness and the reset time is increased by a factor of 10 for integrating and runaway responses. The rate time may also be reduced in some plug flow temperature loops to just equal the thermowell lag time. To mimic the shortcut tuning method, a pseudo integrating process can be selected for the auto tuner or adaptive controller and only the initial ramp used for calculating the tuning settings.

6.1 Batch Control

Some of the more deceptively difficult types of batch control loops involve the simple heat up or neutralization of a batch mixture where there is no crystallization, evaporation, or reaction. If only a hot or cold heat transfer fluid (not both) is available to bring a batch to temperature and there is any reset action in the controller, the batch will always end up beyond its set point. Similarly, if only an acid or base reagent (not both) is available to bring a batch to a pH and there is any reset action in the controller, the batch will always overshoot its set point. The response appears to ramp. Particularly problematic is any residual heat left in the jacket or reagent in the dip tube when the control valve closes.

In this guide the ability to go only in one direction with a manipulated variable will be called "single-ended" control. This case occurs surprisingly often. The solution is a proportional plus derivative controller. The bias should be set to provide zero controller output before the controlled variable (CV) reaches set point since the CV will continue to coast for the duration of the loop dead time after the valve is closed. However, if the product is a validated pharmaceutical where a change in controller structure is problematic, most of the overshoot can be eliminated by setting the reset time a factor of ten or more greater than would be calculated by a Ziegler-Nichols or Lambda tuning rule. An open loop shortcut method is advantageous for these and other batch loops since there is often not enough time or a steady state within the cycle time for a full tuning test.

Rule 43 – **For simple single-ended temperature or pH control, a proportional plus derivative controller is needed to prevent overshoot.** The bias should be set to shut the control valve well before the CV reaches set point.

When the controller output drops to the point where seating friction and changes in the installed characteristic is significant, the controller output can be readily switched to pulse-width modulation to get extra rangeability of the control valve and eliminate the sensitivity, stick-slip, and plugging problems associated with trying to throttle near the seat. The conversion of a percent output to a percent pulse width can be done by a special output card or algorithm. The point for switching to pulsing is about 10% and 20% for sliding stem and ball valves, respectively. If pulse width rather than pulse frequency modulation is used, the valve gain remains constant at its last value.

Rule 44 – **When the controller output drops low, use pulse-width modulation to reduce plugging and to improve the repeatability of the end point.**

6.2 Blending

The accuracy and timing of feeds for a blend is often critical. The set point response for all the feed flow controllers should be identical. Lambda tuning offers this capability. In some cases there is an online analysis and a cascade control system where a master composition loop manipulates the set point of the feed loops in the proper ratio. The dead time in the composition loop heavily depends on the method and location of the analysis, which sets the cycle time and sample transportation delay. Analyzers that require process time or sample systems, create a severely dead time dominant master loop. If fast inline sensors are used, such as Coriolis, microwave, or nuclear magnetic resonance meters, the process response of the master loop is basically the secondary feed loops response set by Lambda tuning. This creates an unstable cascade control system since the basic cascade rule of the master loop being 5 times slower than the secondary loop is violated. The oscillations can be suppressed by slowing down the master loop by detuning and the addition of a filter time. A feedforward of composition set point changes can be added to the master loop output to improve its set point response.

Rule 45 – **Use Lambda tuning to get identical set point responses for the feed flow controllers.** A master composition loop may cause oscillations if it is too fast or has too much dead time from violation of the cascade rule and destabilization, respectively.

A model predictive controller can be set up and tuned to provide tighter composition control. If the changes in feed composition or the desired composition of the blend are slow, a PI controller may suffice. Since the process gain varies with the composition set point, gain scheduling may be beneficial for large changes in the operating point.

6.3 Boilers

The cascade control system where a drum level controller manipulates the set point of a feedwater flow controller must handle the changes in steam use. Normally the level set point is constant. Boiler drum level has an integrating and noisy response with transients that are in the opposite direction of the final response. While the cross-sectional area is relatively large because it is a horizontal vessel, the ramp rate and integrator gain is not exceptionally small because the level range and residence time is smaller compared to a distillate or condensate receiver.

Bubbles and slosh cause measurement noise. An increase in firing rate will increase the bubbles per unit volume in the downcomers, which will push liquid up into the drum and cause its level to swell. A decrease in firing rate will decrease the bubbles per unit volume in the downcomers, which will cause the drum level to shrink. The final level response after the steam generation sufficiently changes the drum liquid inventory is in the opposite direction of the initial response to the change in firing demand. The shrink and swell can be so severe that the boiler may trip on low or high level. The level controller needs to be tuned with mostly proportional action and enough gain to prevent these trips. The reset time will cause overshoot when the inventory changes. While rate action could help deal with shrink and swell, normally measurement noise makes this infeasible. A small filter or velocity limiter on the level measurement can be used to smooth out the signal enough to keep the feedwater valve from responding to level noise passed on as feedwater set point changes. However, a little bit of filter time goes a long way to making an integrating loop less stable and slower to recover from load upsets.

Rule 46 – **Use mostly gain action and judiciously set the signal filter for the drum level controller to prevent shrink and swell from load upsets causing a level trip.** If the measurement is not noisy, rate action may help deal with the transients.

When the controller increases or decreases the feedwater flow, bubbles tend to collapse or expand, respectively, in the drum, which causes a response in the opposite direction of the final response. This action would tend to counteract the shrink and swell from a firing rate change. However, the inverse response for changes in feedwater flow is usually quite minimal in modern systems since the feedwater is preheated and the changes in bubbles in the drum causes a change in density that tends to reduce the change seen by a differential head level sensor. If the controller gain cannot be increased sufficiently to deal with load changes, a transient feedforward signal can be added to the level controller output that is proportional to the negative of the filtered derivative of the firing rate. The classical feedforward signal designed to enforce a mass balance that is proportional to the firing rate should be delayed so as not to interfere with transient correction.

Rule 47 **– If controller tuning is not sufficient to prevent drum level trips, use a transient feedforward to compensate shrink and swell and a delayed classical feedforward to keep the mass in the drum constant.** The transient feedforward and classical feedforward are proportional to the derivative and magnitude of the firing rate, respectively. The transient feedforward gain is negative whereas the classical feedforward gain is positive.

The feedwater flow controller should be tuned first with the Lambda method for an aggressive response that will get through the dead band and resolution of the feedwater valve quickly. The valve is the largest source of dead time in the cascade control system. The process dead time is less than a second. If the controller output can be characterized or the flow controller gain scheduled to compensate for the changes in slope of the installed characteristic of the valve, the Lambda factor can be reduced from 4 to 2.

The level controller should be tuned with a closed loop method for load upsets. Changes in firing rate should be used during the tuning to make sure it can deal with shrink and swell. Make sure the shrink and swell is not interpreted as a closed loop oscillation. Increase the controller gain until an oscillation starts to appear after shrink and swell. A higher level controller gain will increase the size of the flow set point changes, which in turn will help the flow loop get through the valve dead band and resolution faster.

> *Rule 48* – **Use high flow and level controller gains to minimize the dead time from the deadband and resolution of the feedwater valve.** This means some compensation of the feedwater valve nonlinearity is desirable to enable a smaller than usual Lambda factor for the flow controller.

The oxygen controller normally manipulates a bias to the air flow measurement instead of the air flow set point to keep the cross limiting of the fuel and air controls intact. The set point may be optimized as a function of steam load. The oxygen response is dead time dominant unless a large signal filter is used. It is a combination of dead time from the dead band and resolution of the air damper or control valve and the transportation delay of gas from the combustion zone to the oxygen sensor location. Some high temperature zirconium oxide sensors can be located in or near the combustion zone making the process transportation delay less than the dead time from the damper or valve. A sample line to an analyzer house adds a considerable transportation delay.

If the oxygen sensor is in the stack or analyzer house, dead time compensation or model predictive control (MPC) should be used with the dead time automatically adjusted to be inversely proportional to the load. The oxygen and air flow controllers must be tuned for a smooth stable response, where robustness is more important than performance. A larger than normal Lambda factor for PI control and move suppression for MPC is advisable. The oxygen controller should ignore the positive spikes in the oxygen reading from the action of the cross limits for air to lead fuel for an increase in firing demand and air to lag fuel for a decrease in firing demand. The oxygen controller must not chase noise at low load when the oxygen signal gets noisy from poor mixing or at low oxygen levels that are pushing the rangeability limit of the sensor.

Rule 49 – Tune the oxygen controller for a smooth and stable response despite blips from cross limits and noise from low loads.

6.4 Coils and Jackets

Coil and jacket temperature loops are normally secondary loops that must respond to the set point demands from a master temperature loop on a crystallizer, evaporator, fermentor, or reactor. These loops must also respond to disturbances in the coil inlet temperature or flow and depending on the system, changes in the heat transfer coefficient and area. These secondary controllers should be tuned with a closed loop oscillation method before the master controller is tuned.

The velocity is less and the dead time is greater in jacket than a coil but otherwise the considerations are similar. For the rest of this section coils will be discussed, but the conclusions are the same for jackets.

Low coil coolant flow can lead to a high process gain and dead time that can cause instability. If at all possible, the coil flow should be kept constant and the makeup flow manipulated to change the coil inlet temperature. For a recirculation system, the return flow based on a pressure balance should change to match the change in makeup flow.

Rule 50 – **The coil or jacket coolant flow must be constant to prevent oscillations for low cooling demands.** The makeup coolant flow is manipulated.

If the secondary temperature loop controls the coil outlet temperature for a batch reactor, it has an integrator response over part of the operating range. For example, a tuning test with the master loop in manual will show that a decrease in coolant flow will cause an increase in coil outlet temperature but also an increase in reactor temperature, which in turn causes an additional increase in coil outlet temperature. Consequently, the reset time of the coil outlet temperature loop should be increased by to be at least 10 times the loop period.

Rule 51 – **Tune the coil or jacket temperature controller first with a closed loop method and increase the reset time by a factor of 10 or more for outlet loops.**

Coil inlet temperature loops respond faster to upsets in the chilled, cooling tower, or temperature water system. However, these loops do not see the upsets to the heat transfer coefficient or area until after the transportation delay through the recirculation system or at all for a once through system.

When the heat transfer fluid is a vapor such as steam, a secondary loop that controls pressure in the coil would respond well to both changes in supply and heat transfer, if there is always a demand for heat. This pressure loop has an integrating response and should be tuned accordingly with a large reset time for minimal reset action. Controlling pressure is also useful when you can't get a good flow measurement, which might be the case on a low pressure waste steam or heat recovery vapor stream.

Often both heating and cooling is needed and the coil loops have split ranged valves. The process gain, dead time, and time constant is quite different depending on which valve is throttled. The switch over from one valve to another is a severe discontinuity. These loops often oscillate around the split range point. The controller tuning settings must be scheduled based on the valve throttled. Trim valves are used to reduce the effect of stick-slip as one valve opens and another closes. A tempered water system will move most of the impact of these problems away from the sight of the master loop.

Rule 52 – **For split ranged valves, schedule the coil or jacket controller's tuning settings based on the valve throttled and use trim (small) valves to suppress oscillations at the switch over point from heating to cooling.** The transition from a trim to a coarse adjustment valve can be based on split ranged and simultaneous valve position control strategy.

If the coolant flow is manipulated, the use of heat transfer rate as the controlled variable improves the self regulation and linearity of the secondary loop. It also provides a record of the key variable of interest. After all, the master loop temperature really depends on the heat transfer rate, calculated as the product of the coolant mass flow, mass heat capacity and temperature difference across the coils. The inlet temperature should be passed through a dead time block so that a change in coil inlet temperature coincides with a change in coil outlet temperature.

6.5 Compressors

Compressor capacity control loops are incredibly fast when motor speed is manipulated, but generally sluggish when inlet dampers, guide vanes, or recirculation valves are manipulated.

The controlled variable is often compressor discharge pressure. Regardless of the type of manipulated variable or compressor, Lambda tuning should be used to provide a relatively slow and smooth response to avoid upsetting users and interacting with the surge control system. The pressure set point may be slowly optimized based on demand and efficiency to conserve energy.

> *Rule 53* – **For compressor pressure control, use Lambda tuning to provide a gradual and smooth response to production rate upsets and optimization of set points.**

Compressor surge control loops must be incredibly fast to prevent surge. They must have fast valves, scan times, and sensors and be tuned for a fast response. However, even the best feedback loop may not be able to prevent surge or get a compressor out of surge. The precipitous drop in flow occurs in less than 50 milliseconds and the surge cycle period is only 2 to 4 seconds. Consequently, feedforward and open loop back up signals are added to the surge controller output to deal with upsets on a preprogrammed basis. The feedforward signals are usually a summation of feed flows and a "kicker" that increments a surge valve open if the controlled variable, such as suction flow, drops significantly below the surge controller's set point. The "kicker" signal rapidly opens the surge valve and gradually decays away to allow a smooth transition back to feedback control.

> *Rule 54* – **For compressor surge control, use fast devices and scan rates and Lambda tuning to provide a quick but smooth response to production rate upsets.** A feedforward and kicker signal may need to be added to the surge controller output.

6.6 Crystallizers

Crystallizers generally use a cascade control system where the master loop is crystallizer temperature and the secondary loop is the coolant temperature at the inlet to the coil or jacket. The coil inlet temperature must be kept from getting too cold to prevent the formation and the rapid build up of small crystals on the coil surfaces. The frosting of the coil surface degrades the heat transfer coefficient and proper crystal size and growth. A low temperature set point limit must be set and enforced. Lambda tuning is used to provide a rapid but smooth response to set point changes and load upsets in the coolant system.

The run time between defrosting for continuous operation and the cycle time for batch operation depends on the quality of the inlet coil temperature control.

> *Rule 55* – **For crystallizer coil inlet temperature loops, use Lambda tuning to provide tight but smooth control and enforce a low set point limit.** Variability in this loop adversely affects both capacity and quality.

The master temperature controller must be tuned with rate action to deal with the heat transfer lags and crystal growth rate kinetics. A shortcut open loop method is often the most practical. For continuous operation the master set point is fixed, whereas for batch operation the set point is programmed to provide the desired cooling and hence crystal growth rate and size profile during the batch.

> *Rule 56* – **For crystallizer master temperature loops, use the shortcut open loop method and rate action to deal with thermal lags and crystal growth kinetics.**

Some crystallizers are vacuumed controlled to reduce the evaporation of solvent and the temperature of the solution. Here a cascade control system of temperature to pressure is used. The pressure loop should be tuned by a closed loop method to provide an immediate response by the use of mostly gain action before the temperature loop is tuned.

6.7 Distillation Columns

The control scheme seen on many columns because it provides some inherent internal reflux control employs a level controller on the overhead distillate receiver to manipulate the reflux flow and the column temperature controller to manipulate the distillate flow. However, changes in distillate flow only translate to changes in reflux flow that affect the column by action of the level controller. Often the overhead receiver is a horizontal vessel, which means large changes in distillate flow cause small changes in measured level. The integrator gain for the process is very low because the diameter is so large. For this scheme to fulfill its potential, the receiver level controller must be tuned for a very tight response that requires controller gains much higher than normally used. Since gain action amplifies noise, the limit on how high you can go is usually determined by the smoothness of the level measurement.

> *Rule 57* – **When a distillate receiver level controller manipulates reflux flow, use a closed loop tuning method and a judiciously set level measurement filter time to maximize the controller gain.**

The response of the column temperature controller is generally incredibly slow and may take hours or even days to oscillate or reach a new steady state on large columns. Therefore, the shortcut open loop method is used if the closed loop method takes too long. Rate action is essential to compensate for the interactive time constants associated with the concentration response of the trays. A feedforward of feed changes added to the temperature controller output is vital for improved control. If there are temperature controllers on the top and bottom temperatures of the column to control the composition in the distillate and bottoms, model predictive control is essential to deal with the interactions of 2-point composition control, to prevent flooding, and reduce steam use.

Since tray temperature is an inference of composition and large concentration changes often translate to small temperatures, precise temperature measurement and control is important. Resistance temperature detectors (RTDs) and narrow temperature spans are used to ensure accuracy, resolution, and a good signal-to-noise ratio essential for the high rate time settings needed for tight control.

Rule 58 –**For a column temperature controller, use the shortcut open loop tuning method and rate action to compensate for the interactive time constants of the tray concentration response.** Feedforward control, RTD sensors, and narrow temperature calibration spans are critical. For 2-point composition control and optimization, use model predictive control.

There are many other difficulties encountered in the control of columns. For example, the manipulation of steam to control sump level has an inverse response from shrink and swell, and the bottoms temperature controller will interact with a sump level controller, particularly for a thermo-siphon reboiler. Chromatographs used for composition control provide stair cased response at best and an erratic and unreliable signal at their worst.

For Batch columns, the temperature set point is a moving target. Instead of temperature control, a level controller manipulates both the reflux and distillate flow in a ratio set by the batch sequence. This affords the ability for the batch operation to optimize the reflux without upsetting the level controller and the option to go to total reflux.

6.8 Dryers

Dryer temperature control loops suffer from transportation delays and non-representative product temperature measurements. While it is desirable to measure the temperature of the final product, the sensor isn't in good contact with the flowing solids because of voids or buildup of coatings on the thermowell. Consequently a gas temperature measurement is used. Sometimes an inferential moisture measurement (online property estimator) is constructed based on an exit gas temperature subtracted from an inlet gas temperature after it has passed through a dead time and filter block that match the delay and lag time of the dryer. Separate inlet and exit gas temperature loops tend to fight each other. The addition of a feedforward signal based on feed changes is also important. Model predictive control can sort out the interaction, deal with dead time, and provide better feedforward control. If a viewing window can be kept sufficiently clean, an optical pyrometers may provide a more reliable and accurate measurement of the product temperature.

The product is usually much dryer than the specification because of deficiencies in the measurement and control system. By moving the dryer operating point closer to the high moisture limit, moisture can be sold as product and energy costs can be reduced. If the dryer moisture controller manipulates feed, the throughput can be increased. The use of feedforward control and model predictive control, and online property estimators and microwave meters for moisture offer many opportunities for optimization of dryer operation that have been overlooked to date.

Rule 59 – **For dryer temperature controllers use a Lambda tuning method with a small rate setting equal to the thermowell time constant.** If a product temperature measurement is used, the controller gain and rate action may need to be reduced to avoid overreaction to an erratic measurement. For interactions and optimization, use model predictive control. Advanced control and inferential or direct moisture measurements can open the door to more efficient operation.

6.9 Evaporators

The boiling point depends upon the composition and pressure of the contents. Often a temperature loop manipulates the heat input, via a coil or jacket steam flow or pressure controller and a pressure loop manipulates the vapor valve. The boil up and vapor flow is proportional to the heat input. If an analyzer, such as a Coriolis density measurement, is installed in the product discharge or recirculation line, a composition control loop can directly manipulate the heat input. The temperature and pressure measurements become constraints. A model predictive control can increase capacity of the evaporator by maximizing the feed flow without exceeding the high temperature and pressure limits.

Rule 60 – **For evaporator temperature controllers, use a closed loop tuning method with rate action to compensate for thermal lags. Use** model predictive control and a Coriolis meter for composition control and maximization of production rate.

6.10 Extruders

The temperature sensors for the zone temperature controllers are often measuring more the extruder wall or jacket temperature than the polymer temperature. The heat input from these zones act more as a warm blanket and is far exceeded in magnitude by the heat input from the power to the main drive. The zone temperature controllers also interact and are adversely affected by split ranging. Lambda tuning is recommended.

Rule 61 – **For extruder zone temperature control, use a Lambda tuning method to provide a smooth wall temperature.** Make sure that upsets and nonlinearities in the manipulated variables do not affect the wall temperature.

The extruder outlet pressure generally manipulates the extruder speed. If another manipulated variable exists, such as die opening, the speed is freed up for polymer property control. An online property estimator based on residence time and temperature rise may prove valuable here.

The extruder pressure loop must have a fast and smooth response. The sensor and manipulated variable response time and scan rates must be as fast as possible. If die opening is manipulated, hydraulic actuators should be used. If the loop cannot be made fast enough, a feedforward signal based on extruder or sheet speed should be added to the pressure controller output to help it deal with production rate changes.

Rule 62 – **For extruder zone temperature control, use a Lambda tuning method to provide a smooth wall temperature.** Make sure that upsets and nonlinearities in the manipulated variables do not affect the wall temperature.

The extruder outlet pressure generally manipulates the extruder speed. If another manipulated variable exists, such as die opening, the speed is freed up for polymer property control. An online property estimator based on residence time and temperature rise may prove valuable here.

The extruder pressure loop must have a fast and smooth response. The sensor and manipulated variable response time and scan rates must be as fast as possible. If die opening is manipulated, hydraulic actuators should be used. If the loop cannot be made fast enough, a feedforward signal based on extruder or sheet speed should be added to the pressure controller output to help it deal with production rate changes.

Rule 63 – **For extruder pressure controller, use a Lambda tuning method, with fast hardware and scan rates to provide a fast and smooth response to load upsets.** If this loop is not fast enough, a feedforward signal of speed changes is useful for minimizing the variability from changes in the production rate. An online property estimator that manipulates extruder speed may greatly improve product quality.

6.11 Fermentors

Cell concentration, dissolved oxygen or carbon dioxide, pH, and temperature are the important loops for fermentor and bioreactor control. The upsets are incredibly slow and mostly involve the transitions from exceptionally low demands for the initial low cell concentration to the high demands in the exponential phase of high cell growth and product formation rates and finally back to a low demand in the stationary phase as the cells mature. Split ranged manipulated variables are used to deal with extreme variations in demand for carbon dioxide or oxygen, cooling, nutrients, and reagents. In general, a temperature water system is used to ensure the jacket temperature is moderated. Shocks to the cells from localized extremes of any operating conditions must be avoided.

It is critical that the cells do not see any extremes so the controllers are tuned for a gradual and very slow response. Any tuning tests must be short and involve very small excursions for short time intervals, which can be achieved by a shortcut tuning method.

Rule 64 – **For fermentor or bioreactor control, use a shortcut tuning method with a greatly reduced gain and decreased rate setting to provide a smooth and very slow response.**

The loops interact and the split ranging of manipulated variables changes the process dynamics but the detuning and the slow nature of the upsets results in straight lines of the controlled variable throughout the batch. The effect of batch conditions instead show up in the trends of the controller outputs.

The use of mass spectrometers for vapor analysis and the calculation of oxygen uptake rates and carbon dioxide production rates can be used to create online property estimators. These estimators along with strategies to optimize a profile during a batch to improve yield or reduce cycle time will increase the need for better tuning and model predictive control.

6.12 Heat Exchangers

The response of a temperature control loop is dominated by a transportation delay and the thermal lags of the thermowell. If coolant flow is throttled, low velocities can cause a large dead time and a high process gain and greater surface fouling problems. If the coolant valve is upstream of the exchanger, water can flash under some conditions. In general, the manipulation of a bypass flow instead of the coolant flow greatly reduces the dead time and improves the controllability. In general, Lambda tuning is sufficient unless the exchanger is in a recirculation line of a vessel, which creates an integrating response. In this case, a closed loop method with the reset time set at least 10 times the loop period and a rate setting equal to the thermowell lag is advisable.

Rule 65 – **For heat exchanger temperature controllers, a Lambda Tuning method with a rate time equal to the thermowell lag time is sufficient. However an exchanger in a recirculation line for a vessel should use a closed loop method with a 10 times larger reset time.** Normally recirculation temperature controllers are secondary loops whose set point is manipulated by a master reactor temperature controller.

6.13 Neutralizers

The pH measurement offers an incredible sensitivity and rangeability that translate to exceptional mixing, piping design, control valve resolution, and controller tuning requirements. A steep titration curve slope amplifies what may normally be negligible amount of valve stick-slip and feed upsets. The nonlinearity can require orders of magnitude change in controller gain depending upon the operating point. The scheduling of controller gain as a function of pH or the translation of the controlled variable from pH to percent reagent demand based on the titration curve is highly recommended unless the set point is on the relatively flat portion of the titration curve.

Rule 66 – **For a pH set point on the steep portion of a titration curve, schedule the controller gain or translate the controlled variable from pH to percent reagent demand based on the curve.**

Most pH loops on even well-mixed vessels exhibit considerable dead time because of reagent injection delays caused by small reagent flows. If the pH is driven from a knee to the steeper portion of the titration curve, the acceleration of the response can make the loop appear to lack self-regulation. The use of rate action seems to help pH loops on vessels deal with the nonlinearity of the titration curve and pseudo runaway response.

Rule 67 – **Use incredibly precise reagent valves, specially designed reagent injection systems, and a shortcut open loop tuning method with considerable rate action for vessel pH control.** The measurement will need to be filtered to keep fluctuations of the controller output within the resolution of the control valve.

6.14 Reactors

The principal concern with reactors is the exponential dependence of the reaction rate on temperature and the possible development of a runaway (open loop unstable) response. Plug flow or catalyst bed reactors have a significant dead time determined by the transportation delay and an axial temperature profile. Back mixed (stirred) reactors have a small dead time set by the mixing delay and are designed to have a uniform temperature throughout the mixture. A closed loop tuning method and full rate action is recommended for both. For severe exothermic reactors, such as those used in polymer production, proportional plus derivative (PD) controllers are used because reset action is considered too risky. If a PID controller is used on an exothermic reaction with a non-self regulating response, the reset time should be increased to be at least 10 times the loop period.

Rule 68 – **For reactor temperature control, use a closed loop method with full rate action.** For exothermic reactors, the reset time should be increased by a factor of ten or more. For a severe runaway response, reset action should not be used.

When the product or byproduct is a vapor, a part of the vapor flow is often condensed and returned (refluxed) to the reactor to improve the purity of the vapor stream. The exit temperature or heat removal rate of the overhead partial condenser is controlled by manipulating the cooling water flow or a small nitrogen flow to partially blanket the condenser's heat transfer surface. The condenser loop will interact with reactor temperature and pressure loops by changing the reflux and vapor flow rates, respectively.

> *Rule 69* – Partial condenser temperature loops should use a Lambda tuning method for a smooth response with a rate time set equal to the thermowell time lag.

6.15 Remote Cascade

Control loops that operate in a remote cascade (RCAS) mode where the set point is manipulated by model predictive control (MPC) should use a Lambda tuning method that provides a consistent and closed loop time constant. MPC is sensitive to a dead time mismatch between its model and the actual response of its controlled variables to the manipulated variables. A faster response of an RCAS loop than what was identified during testing is actually more of a problem than a slower response, because MPC like the Smith Predictor is more oscillatory for an overestimate of the dead time. The existence of an offset also confuses MPC. If reset action cannot be used in sufficient amount to sufficiently eliminate offsets in the set point response because the RCAS loop response is not much faster than the MPC response, the offset can be added as a disturbance variable to the MPC.

Rule 70 – **Loops that operate in the remote cascade (RCAS) mode, should use a Lambda tuning method to provide a consistent and smooth set point response with minimal offset.** A fixed non-oscillatory set point response with a negligible offset is critical for minimizing model mismatch in MPC.

6.16 Sheets and Webs

Sheet production rate control is incredibly fast. Each line roller speed must be properly ratioed to the manipulated master roller speed. Lambda tuning and very fast scan rates are essential. Breaks in the sheet can cause an enormous waste of raw materials and loss of production, particularly for large and fast sheet lines.

> ***Rule 71*** – For sheet production rate control, use a Lambda tuning method for a smooth response with properly ratioed roller speeds and a very fast scan rate.

Thickness control is a loop dominated by the transportation delay from the die to the density or thickness gage. The dead time in a Smith Predictor or MPC must be updated based on sheet speed. The control of the thickness along the length of the sheet is called machine direction or MD control. The control of thickness across the sheet is called cross direction or CD control. For MD control, the manipulated variable is a takeup roller speed that is ratioed to the master roller speed used for production rate control. For CD control, the manipulated variables are a series of die bolts with actuators or heaters. To date, MD and CD control has been predominantly done by decoupled Smith Predictors. Exceptionally large and fast matrix algorithms has enabled the migration of MD control and some CD control to MPC for optimization. However, the controlled variables for wide sheets that have a 100 or more CD controlled and manipulated variables still require some consolidation of variables to fit within today's MPC.

Rule 72 – For PI control of sheet MD and CD thickness, use a Lambda tuning method and decoupled Smith Predictors with a dead time based on line speed.

7.0 – Adaptive Control

7.1 Learning the Terrain

The next generation of adaptive controllers will identify the process dynamics and allow the user to pick the most appropriate tuning rule for their process. By identifying and remembering the process terrain, the controller does not have to rehash old information and can concentrate on changes. Knowledge of the changes in the open loop gain, dead time, and time constant enables recognition of changes in raw materials, equipment, conditions, and the response of instrumentation and control valves that are important for plant performance. For example, a decrease in valve gain can be caused by a partially plugged filter and an increase in loop dead time can be caused by a fouled thermowell.

Rule 73 – **Use knowledge of the changes in the process dynamics to diagnose equipment and instrumentation problems.** The changes in process gain often reflect changes in the composition, pressure, and temperature of streams, whereas changes in the process dead time often reflect changes in the fouling of heat transfer and sensor surfaces.

7.2 Watching but Not Waiting

If the process dynamics from previous excursions in different operating regions of the actual plant or a virtual plant (dynamic process model) are stored as a function of key process variable or manipulated variable, the controller tuning can be scheduled. Figure 23 shows 4 different pH regions with vastly different process gains that reflect the nonsymmetrical shape of the titration curve for a mixture of acetic acid and ammonia with a dash of sulfuric acid and the absorption of carbon dioxide from the air. This allows the adaptive pH controller to take preemptive action when it moves to a new operating point. Previously, adaptive controllers had to see step changes or oscillations even though the topography had not changed. These controllers were always playing catch-up.

Rule 74 – Schedule the controller tuning settings based on knowledge gained from previous excursions into different operating regions of an actual or virtual plant.

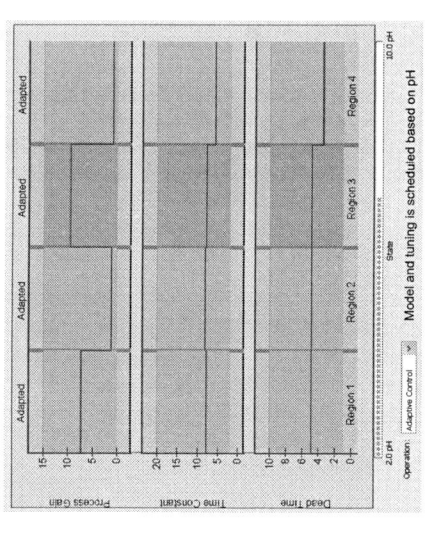

Figure 23 – Process Models and Tuning Settings Scheduled by an Adaptive pH Controller

ADAPTIVE CONTROL

7.3 Shifting into High Gear

Most processes are not operating at the most optimum set point primarily because of a lack of process knowledge and excessive process variability. The set point chosen by operators are based more on tradition and opinions rather than an understanding of more efficient operating points. A zoom-in on the titration curve in the control region in Figure 24 shows the opportunity for the neutralization of an acidic influent by lowering the set point closer to the low pH limit. If the influent becomes basic, the set point would be switched to be closer to the high pH limit. In this case, stick-slip and feed upsets (oscillations on the X axis) have less impact at the more optimum set point.

> *Rule 75* – Use process knowledge and an adaptive controller to move the process to a more efficient operating point.

Figure 24 – Opportunity for an Adaptive pH Controller to Reduce Reagent Use

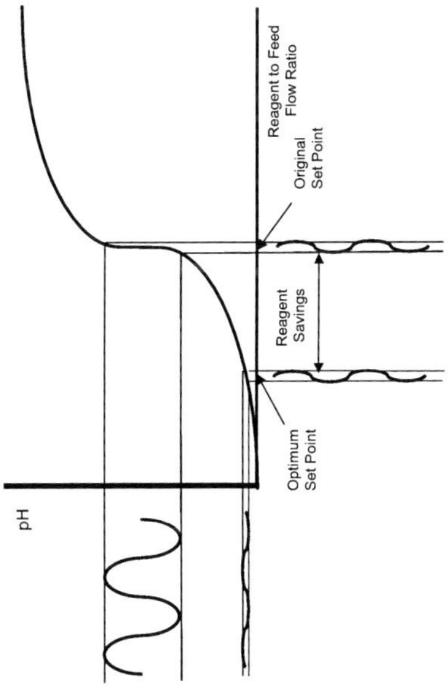

ADAPTIVE CONTROL

7.4 Back to the Future

This new generation of an adaptive controller allows all PID loops to run in the adaptive mode and provides process model parameters that are saved in a data historian and analyzed for changes in the plant, sensors, and valves. The information on changes in the process model may be directly used to monitor loop performance and to provide more intelligent diagnostics. The models can provide the dynamics for simulations and identify candidates for feedforward control and advanced control techniques. For example, loops dominated by a dead time or exhibiting disturbance models for multiple variables, are prime candidates for model predictive control. Feedforward models of compositions could be used for dynamic online property estimators and loops dominated by a single large time constant could benefit from fuzzy logic control. The dynamic process models in general can be used to create or adapt real-time simulations for prototyping new control strategies, exploring "what-if" scenarios, and training operators. Process gains that decrease or time constants that increase with feed totals are ripe for real-time optimization of the run time between defrosting or cleaning and catalyst reactivation or replacement. The beauty of this route is the models and tuning settings are available from the adaptive controller for a higher level of control by a better knowledge of the topology.

Online process performance indicators with economic factors, such as the cost of excess reagents, reactants, or energy, can provide the knowledge, motivation, and justification to invest in the use of new technologies to find and exploit more efficient operating points. Figure 25 shows how an online indication of excess reagent cost was greatly reduced by the use of an adaptive pH controller to get to a more efficient set point.

Rule 76 – Use online performance indicators to provide the knowledge, motivation, and justification to install and keep improved control systems in their highest mode.

If you understand and practice everything in this pocket guide, you might become the next CEO of your company. If this is not realistic, maybe you will be given stock options for every loop that is tuned. If this is a pipe dream, the improvement in loop performance might lead to a promotion. Well, maybe you will get a gift certificate to Burger King.

Seriously, I wish you all the best of luck in tuning loops.

Figure 25 – The Reagent Savings Indicated Online by an Adaptive pH Controller

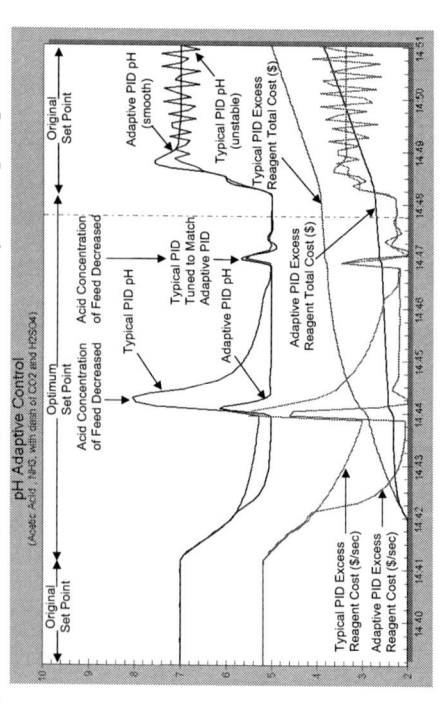

Appendix A—Technical Terms in Process Control that Are Used Interchangeably

Attenuation – Filtering – Smoothing

Backlash – Dead Band

Delay Time – Dead Time*

Derivative Action – Rate Action**

Derivative Time – Rate Time**

Digital Filter – Measurement Filter – Process Variable Filter – Signal Filter

Fouled – Coated

Integral Action – Reset Action**

Integral Time – Reset Time**

Lag Time – Time Constant*

Process Variable Gain – Process Gain*

Sticktion – Resolution - Sensitivity

Scan Time – Cycle Time – Execution Time

Steady State Gain – Static Gain*

System Dead Time – Total Loop Dead Time*

Upset – Disturbance – Load Change

* Dynamic Model Parameters

** Controller Tuning Parameters

Appendix B—For Math Lovers Only

The simplified Lambda tuning equation for the controller gain can be derived from the original equations for Lambda tuning. The original equations are useful for designing loops to have the same closed loop time constant for blending operations and model predictive control (MPC), particularly when there are multiple production lines. In these cases, Equation B-1 is solved for the Lambda factor (λ_f) that gives the desired closed loop time constant (τ_c).

Original equations:

$$\tau_c = \lambda_f * \tau_o \tag{B-1}$$

$$T_i = T_{98} / 4 \tag{B-2}$$

$$K_c = T_i / [K_p * (\tau_c + \tau_{do})] \tag{B-3}$$

Time to 98% of final response is 4 time constants plus the dead time:

$$T_{98} = 4 * \tau_o + \tau_{do} \tag{B-4}$$

Substituting Equation B-4 into Equation B-2 shows the dependence of the integral time on the open loop time constant and the observed dead time:

$$T_i = \tau_o + \tau_{do}/4 \qquad (B-5)$$

Substituting Equations B-1 for the closed loop time constant (τ_c) into the denominator and B-5 for the open loop time constant corrected for dead time into the numerator of B-3:

$$K_c = (\tau_o + \tau_{do}/4)/[K_o * (\lambda_f * \tau_o + \tau_{do})] \qquad (B-6)$$

Factoring out the Lambda factor in the denominator:

$$K_c = (\tau_o + \tau_{do}/4)/[K_o * \lambda_f * (\tau_o + \tau_{do}/\lambda_f)] \qquad (B-7)$$

Substituting the Lambda factor for robustness ($\lambda_f = 4$) for the devisor of τ_d in the denominator of B-7:

$$K_c = (\tau_o + \tau_{do}/4)/[K_o * \lambda_f * (\tau_o + \tau_{do}/4)] \qquad (B-8)$$

Canceling out the common term

$$(\tau_o + \tau_{do}/4)$$

in the numerator and denominator of B-8:

$$K_c = 1 / [K_o * \lambda_f] \qquad (B-9)$$

Substituting the definition of the open loop gain ($K_o = \Delta\%PV / \Delta\%CO$) into B-9, we end up with the simplified equation for the controller gain shown in section 2.5:

$$K_c = (\Delta\%CO / \Delta\%PV) / \lambda_f \qquad (B-10)$$

where:

$\Delta\%CO$ = change in controller output (%)
$\Delta\%PV$ = change in process variable (%)
K_c = controller gain setting
K_o = open loop gain ($\Delta\%PV / \Delta\%CO$)
λ_f = Lambda tuning factor
T_i = controller integral time setting (sec/repeat)
τ_c = closed loop time constant (e.g., set point response time constant) (sec)
τ_{do} = observed dead time (sec)
τ_o = open loop time constant (e.g., largest process time constant) (sec)

Note that $\Delta\%PV$ is the same as ΔCV since the controlled variable is the process variable expressed as a percent of the measurement span corresponding to the controller scale.

Appendix C—An Integral Part of Tuning

The integral time setting presented at the end of Chapter 2 is important for the secondary loops of cascade control systems and for integrating and runaway processes in general. For primary or single loops on self-regulating processes, the factor for the closed loop tuning method applied to the loop period to get the best integral time setting varies from about 1/8 to 1.0 for self-regulating processes, which corresponds to a pure dead time and lag response, respectively. Equation C-1 has been shown to be effective at estimating an integral factor of ¼ to 1 for industrial systems where pure dead time processes are rare.

$$k_i = 1/\{[(4 * \tau_d/T_o) - 1)]^2 * 3 + 1\} \quad (C\text{-}1)$$

$$T_i = k_i * T_o \quad (C\text{-}2)$$

If $\tau_d < T_o / 3$ then (otherwise $k_d = 0$)

$$k_d = 0.1 * k_i \quad (C\text{-}3)$$

$$T_d = k_d * T_o \quad (C\text{-}4)$$

For the shortcut method, a process gain of one is assumed so that the process time constant is the inverse of the pseudo integrator gain. The shortcut dead time factor is then the closed loop period factor multiplied by the loop period divided by the dead time.

$$c_i = k_i * (T_o/\tau_d) \tag{C-5}$$

$$c_d = k_d * (T_o/\tau_d) \tag{C-6}$$

The following equations have been used to approximate the loop period in various applications for self-regulating processes:

$$T_o = 1.4 * \{[(1 - e^{-(3*\tau_d/\tau_o)})/(1 - e^{-(\tau_d/\tau_o)})] + 1\} * \tau_d \tag{C-7}$$

$$T_o = 1.4 * \{[(\Delta PV_1 + \Delta PV_2 + \Delta PV_3)/\Delta PV_1] + 1\} * \tau_d \tag{C-8}$$

$$T_o = 1.4 * \{1 + [(\tau_d/(\tau_d + \tau_o)]^{0.65} + 1\} * \tau_d \tag{C-9}$$

where:

c_i = integral time factor applied to total dead time in shortcut method

c_d	=	derivative time factor applied to total dead time in shortcut method
k_i	=	integral time factor applied to loop period in closed loop method
k_d	=	derivative time factor applied to loop period in closed loop method
ΔPV_1	=	change in the process variable within the 1st dead time interval (e.u.)
ΔPV_2	=	change in the process variable within the 2nd dead time interval (e.u.)
ΔPV_3	=	change in the process variable within the 3rd dead time interval (e.u.)
T_i	=	controller integral (reset) time setting (sec/repeat)
T_d	=	controller derivative (rate) time setting (sec)
T_o	=	quarter amplitude loop period (sec)
τ_d	=	total loop dead time (sec)
τ_o	=	open loop time constant (e.g., largest process time constant) (sec)

Appendix D—Closed Loop Time Constant

The time constant used in this guide to characterize a process response is an open loop time constant, which is the time constant for the response of the controlled variable to a change in controller output with the controller in manual. It is increasingly popular to specify a closed loop time constant, which is the response of the controlled variable to a change in controller set point with the PID algorithm active (Auto, Cas, or Rcas modes). For loops whose set point is manipulated by a batch, blend, model predictive control, or ratio control system, a closed loop time constant is an effective way to specify a consistent set point response that is essential for the performance of these systems.

Lambda tuning provides a closed loop time constant (τ_c) that is approximately equal to the open loop time constant (τ_o) multiplied by the Lambda tuning factor (λ_f).

$$\tau_c = \lambda_f * \tau_o \qquad (D\text{-}1)$$

The closed loop time constant (τ_c) can be estimated for any tuning method as the integral time (T_i) (i.e., reset time) setting multiplied by the product of the controller gain (K_c) and the dimensionless open loop gain (K_o) defined in Section 1.6.

$$\tau_c = K_c * K_o * T_i \qquad (D\text{-}2)$$

Index

A
abrupt responses 19, 27
accuracy
    auto tuner 1
    measurement (see measurement resolution)
    tuning 1
action
    control 2–6
    direct 2–6
    fail-open 2–6
    gain 7, 14, 16, 22, 96–97, 101–102, 105
    on-off 99, 101, 107
    process 2–6
    rate 8, 14, 38, 52, 95, 97, 101–102, 105
    reset 7, 14, 16, 20, 22, 24, 51–52, 85, 88–89, 94, 96–97
    reverse 3–6
    valve 2–6
actual valve position 57
actuator 66, 71
adaptive control 151
advanced control tuning 8, 16, 44
algorithm
    error-squared 20, 52
    PID 9, 47
    update time 105, 110–111
aliasing signal 105
analog output block 4
analog/digital converter (A/D) 27, 29, 63, 76
analyzer
    cycle time 27
attenuation (see filter)
auto mode 6
auto tuner
    accuracy 1
    identification of dead band 85
    integrator gain 40
    pretest 87
    relay method 85–86
automatic mode 1, 6

B
back-calculate
    function blocks 4, 7
backlash (see dead band)
ball valve 66, 71
batch control tuning 8, 16, 44, 115
blending 118
    (see also pipeline and static mixer)
boilers 119
boiling point 137
bump 43, 51, 53, 60
butterfly valve 12, 36, 66, 71

C
calibration 64
calibration span 11, 27, 37
cas (cascade) mode 6
cascade control 3, 8–9, 16

cascade control tuning 44, 48, 95–98
characteristic
   valve 12, 22, 31, 36–37, 71, 73, 95
characterization
   signal 71, 73, 94
chatter 27–28
closed loop 6–7, 44, 47–48, 51–57, 58
coated 44, 78, 88
coating of sensor 63–64, 85
coils 125
   inlet temperature 126
column control tuning 14, 19, 21–22, 24, 36, 53, 92
column control valve 70
compression 105, 110–111
compressor 128
   control tuning 37, 39
   control valve 66, 70
   surge control 129
concentration control tuning 27, 29, 38, 85, 96
control
   action 2–6
   bump 43, 51, 53, 60
   cascade 3, 6, 8–9, 16, 19
   disturbance (see also upset) 30, 88, 92, 95
   error 7–8, 11, 28
   feedback 2, 33, 88
   feedforward 52, 92–94, 96
   interaction 16, 20, 44, 52–53, 58, 62, 95, 97, 99–100
   limit cycle 55
   material balance 21
   offset 12, 24, 96, 109
   oscillations 16, 21–25, 34, 48–55, 58, 70, 85, 88
   overshoot 8, 14, 24–25, 101
   performance 28, 33, 92, 95, 111
   residence time 35, 37
   resonance (loops) 99, 101
   set point 8, 14, 16–18, 20, 23
   split-ranged 14
   stability 22, 24
   upset (see also disturbance) 30, 52–53, 55, 60, 92, 95–96, 99, 101, 105, 107
   variability 1–2, 16, 19, 27, 107
controlled variable 11, 37, 90
controller output
   function blocks 4, 12
   PID 12, 40, 51, 55, 57, 60
conveyor control tuning 19, 24, 58
crystallizer control tuning 14, 21
crystallizers 130
current-to-pneumatic transducer (I/P) 4
cycle time (see scan time)
cycle time analyzer 27

D

dead band 19, 47, 53, 56, 64, 73-74

valve 2, 19–20, 29, 33, 47–48, 51, 53, 55, 60, 64, 71, 73
dead time
of valve 47, 53, 55–56, 58, 60, 71, 73
dead time or time delay response 30–34, 53–57, 88–90, 92, 95, 97
dead time to time constant ratio 16, 35
dead-time compensator tuning 88–92, 108
dead-time dominant response 16, 19, 35–36, 52, 58, 101, 108
delay
from transportation 33, 88–89
delay time (see dead time)
derivative
action (see rate action)
mode 8, 10
time (see rate time)
desuperheater control tuning 19, 36, 38, 58
digital filter (see measurement filter)
digital positioner 64, 71
direct action 2–6
direct digital control (DDC) 7
distillation 132
(see also column control)
disturbance control (see also upset) 30, 88, 92, 95
drift sensor 63–64
dryers 135

E

dynamic classes
valve 66, 69

error
control 8, 28
error control 7–8, 11, 28
error-squared algorithm 20, 52
evaporator control tuning 14, 21, 96
evaporators 137
execution time (see scan time)
exothermic 145
extruder outlet pressure 139
extruders 138

F

faceplate
PID 14
fail-open
valve 4–6
falter or hesitate response 16, 101
feedback control 2, 33, 88
feedforward control 52, 92–94, 96
feedwater flow controller 122
fermentors 140
fieldbus function blocks 4, 9, 98
fieldbus mode 6–7
filter 8, 12, 19, 27, 51, 55, 60, 94–95, 97
flow (feed) ratio 36–40
flow control tuning 19, 38, 52, 85, 89, 95

INDEX 173

fluctuations or waves
pressure 99–100
fouled 88 (see also coated)
function blocks
   back-calculate 4, 7
   controller output 12
   fieldbus 4, 9, 98
   PID 4

G

gain action 7, 14, 16, 22, 96–97, 101–102, 105
gain tuning 7–11, 19–24, 35, 47–48, 51–52, 55, 57, 60–61, 102, 109–110
general-purpose method tuning 44–48

H

heat exchanger control tuning 19, 36, 38
heat exchangers 142
heat transfer 127
historian
   compression 105
   update time 105

I

identification of dead band 85
increase-to-close valve 3–6
inferential moisture measurement 135–136
initialization manual (IMan) mode 7
integral action (see reset action)
integral mode 7, 10–11
integrating response 47, 108

integrator gain
   for auto tuner 40
integrator response 39–41
interaction 95, 97
   control 16, 20, 44, 52–53, 58, 62, 99–100
inverse response 27, 52–53, 61, 90, 92, 94

J

jackets 125

K

"kicker" algorithm 129–130

L

lag
   thermowell 29
   time (see also time constant)
Lambda factor 122
Lambda tuning 35, 38, 57–62, 96–97
level
   control valve 63, 70–71
   controller 122
   switches 99, 105, 107
   tuning 19–21, 24–25, 29, 44, 51–52
limit cycle 55
limits
   output 51, 97–98
   set point range 97–98
   set point velocity 8, 19, 52
load change (see disturbance)
local mode 6
local override mode 6

loop
   closed 6–7, 44, 47–48, 55, 58
   master (also known as primary) 8, 56
   open 6, 47, 53, 58
   primary (also known as master) 8, 56
   secondary (also known as slave) 8, 48, 95–98
   slave (also known as secondary) 8, 48, 95–98
   troubleshooting 99–111

M

maintenance
   of valve 64, 71–72
manipulated variable 3, 9, 11, 22, 28, 31, 36, 39–40
manual (Man) mode 6
master loop (also known as primary) 8, 56, 95–98
master temperature controller 131
material balance control 21
measurement filter 29, 55, 60, 133
measurement resolution 27–28, 110–111
methods
   tuning 1, 16
mode
   auto 6
   automatic 1, 6
   cascade (Cas) 6
   derivative 8, 10
   direct digital control (DDC) 7
   fieldbus 6–7
   initialization manual (IMan) 7
   integral 7, 10–11
   local (see auto) 6
   local override (LO) 6
   manual (Man) 6
   operational 6
   output tracking 6
   proportional 7, 9–12
   remote (see cascade) 6
   remote cascade (RCas) 7
   remote output (ROut) 7
   remote set point (RSP) 6
   supervisory 7
model predictive control (MPC) 28, 124

N

neutralizers 143
noise signal 19–20, 27, 30, 42, 44, 51, 55, 60, 63, 70
noisy measurements 81–84
nonlinearity response 1, 36, 47, 58, 94–95

O

offset control 12, 24, 96, 109
on-off action 99, 101, 107
open loop 6, 47, 53, 58
open-loop gain response 34–42, 57, 60–61
operational mode 6
oscillations 16, 21, 24, 34, 48–55, 58, 70, 85, 88
   control and response 16
output limits 51, 97–98
output tracking 6
overshoot control 8, 14, 24–25, 101

oxygen controller 123
oxygen sensor 124

P

patterns of response 92
performance
    control and tuning 28, 33, 92, 95, 111
pH
    control tuning 27, 29, 31, 36, 55
    control valve 70, 88
    measurement 143
PID
    algorithm 9, 47
    controller output 12, 40, 51, 55, 57, 60
    faceplate 14
    function blocks 4
    scan time 11, 28–29, 45–46, 53, 63, 85, 95, 97, 107, 110
pipeline control tuning 19, 24, 27, 38, 58, 89
positioner
    digital 64, 71
    pneumatic 64, 70
    smart (see digital positioner)
    valve 4, 64–71
power spectrum signal 105–106
pressure
    fluctuations or waves 99–100
    control tuning 16, 24, 35–37, 44, 48, 51, 85, 96, 100
    control valve 66, 71
    waves 99, 101
pretest
    auto tuner 87
primary loop (also known as master loop) 8, 56, 95–98
process
    action 2–6
    dynamics 152
    gain 36-39, 89–90, 95, 114, 119, 125, 127, 142, 151, 156
    variable 2–3, 8–9, 11, 14, 16, 19, 27, 31, 36, 40, 51, 57–58, 60, 88, 90
    variable filter (see measurement filter)
    variable gain (see process gain)
proportional
    action (see gain)
    band tuning 9, 11, 20
    mode 7, 9–12
pulp and paper control tuning 20
pump control tuning 37, 39

Q

quarter-amplitude response 21–22

R

ramp rate 39–41, 53–55, 107, 109–110
rate
    action 8, 14, 27, 38, 52, 95, 97, 101–102, 105
    time to dead time ratio 52
    tuning 8, 10, 25–27, 47, 58

ratio
 dead time to time
  constant 16, 35, 108
 flow (feed) 36–40
 rate time to dead time 52
 reset time to oscillation
  period 24, 48, 52
 signal-to-noise 63
reactor control tuning 14, 16,
 21–22, 44, 48, 51, 96, 109
reactors 145
reagent demand control 143-
 144
relay method
 auto tuner 85–86
remote
 cascade mode 7, 146
 mode (see cascade) 6
 output (ROut) mode 7
 set point (RSP) 6, 98
reset
 action 7, 14, 16, 20, 22,
  24, 51–52, 85, 88–89,
  94, 96–97
 time to oscillation period
  ratio 24, 48, 52
 tuning 10–11, 14, 22–35,
  47
residence
 time control 35, 37
 temperature detectors 63
resolution
 measurement 27–28,
  110–111
 valve 27–29, 64, 71
response
 abrupt 19, 27
 analyzers 27

chatter 28
dead time or time delay
 53–57, 88–90, 92, 95,
 97
dead-time dominant 16,
 18–19, 35–36, 52, 58,
 101, 108
falter or hesitate 16, 101
integrating 47, 108
integrator 39–41
inverse 27, 52–53, 61, 90,
 92, 94
nonlinearity 1, 36, 47, 58,
 94–95
open-loop gain 57, 60–61
oscillations 70
patterns 92
quarter-amplitude 21–22
runaway 47–48, 108
self-regulating 40
sensor 63
set point 44, 48, 51–52,
 55, 60
smooth 19–20, 35, 88, 96
staircase 25, 101
steady-state gain 90
time constant 8, 16, 20,
 27, 37, 40, 89–90, 95,
 97
time to steady state 58–
 62, 66–69
trajectory 14, 29
reversal
 of signal 3–6
rotary valve 66, 70–71
runaway response 47–48, 108
response
               runaway 22

INDEX   **177**

S

scan time 28–29, 45–46, 53
  PID 11, 63, 85, 95, 97, 107, 110
scheduled tuning 143–144, 152
secondary (also known as slave) 95–98
secondary loop (also known as slave) 8, 48, 95–98
self-regulating response 40
sensitivity (see resolution)
  of valve (see valve resolution)
  to measurement (see measurement resolution)
sensor
  coating 63–64, 85
  drift 63–64
  response 63
sequence control tuning 19, 44
set point
  control 8, 14, 16–18, 20, 23
  range limits 97–98
  response 44, 48, 51–52, 55, 60
  velocity 8, 19
  velocity limits 52
settings for tuning 43, 45, 51, 53
shaft of valve 66, 68, 71, 73
sheet (web) control tuning 19–20, 24, 58
sheets 147
shortcut tuning 85

shrink and swell 120-122, 134
signal
  aliasing 105
  analog to digital converter (A/D) 27, 29, 63
  current-to-pneumatic transducer (I/P) 4
  filter 8, 12, 19, 27, 51, 55, 60, 94–95, 97 (see also measurement filter)
  noise 19–20, 27, 30, 42, 44, 51, 55, 60, 63, 70
  power spectrum 105–106
signal characterization 71, 73, 94
signal reversal 3–6
signal-to-noise ratio 63
single-ended control 116
size
  step 43, 53, 58, 66, 69, 100, 110–111
slave loop (also known as secondary) 8, 48, 70, 95–98
sliding stem (globe) valve 71
smart positioner (see digital positioner)
smart transmitter 63–64
smooth response 19–20, 27, 35, 88, 96
smoothing (see filter)
span
  calibration 11, 27, 37
specification
  valve 66
spinning (fibers) control tuning 19

split range valve 70
split-ranged control 14
stability 22, 24
staircase response 25, 101
startup control tuning 16
static gain (see steady-state gains)
static mixer control tuning 19, 36, 38
steady-state gains 30–42, 90
steam traps 99, 101
step size 43, 53, 58, 66, 69, 100, 110–111
stick-slip 74–77
   valve 31, 64, 102, 107
sticktion (see resolution)
stroking time (see valve response time)
supervisory control tuning 7
supervisory mode 7
system dead time (see total loop dead time)

T

temperature
   control tuning 14, 24, 27–29, 35, 38, 48, 52
   control valve 70–71
   detectors 63
   loop controls 126
thermowell lag 29
time constant response 8, 16, 20, 27, 30–37, 40, 89–90, 95, 97
time delay (see dead time)
time to steady state response 58–62
total loop dead time 42, 62, 88

trajectory of response 14, 29
transmitter
   smart 63–64
transportation delay 33, 88–89
trend recording
   compression 105, 111
   update time 105, 111
troubleshooting loop 99–111
tuning
   accuracy 1
   advanced control 8, 16, 44
   batch control 8, 16, 44
   cascade control 44, 48, 95–98
   closed loop method 51–57
   column control 14, 19, 21–22, 24, 36, 53, 92
   compressor control 37, 39
   concentration control 27, 29, 38, 85, 96
   conveyor control 19, 24, 58
   crystallizer control 14, 21
   dead-time compensator 88–92, 108
   desuperheater 36
   desuperheater control 19, 38, 58
   evaporator control 14, 21, 96
   flow control 19, 38, 52, 85, 89, 95

INDEX 179

gain 35, 47–48, 51–52, 55, 57, 60–61, 102, 109–110
general-purpose method 44–48
heat exchanger control 19, 36, 38
Lambda 35, 38, 57–62, 96–97
level 19–21, 24–25, 29, 44, 51–52
pH control 27, 29, 31, 36, 55
pipeline control 19, 24, 27, 38, 58, 89
pressure control 16, 24, 35–37, 44, 48, 51, 85, 96, 100
proportional band 9, 11, 20
pulp and paper control 20
pump control 37, 39
rate 8, 10, 47, 58
reactor control 14, 16, 21–22, 44, 48, 51, 96, 109
requirements 113–115
reset 10–11, 14, 22–35, 47
sequence control 19, 44
settings 43, 45, 51, 53
sheet (web) control 19–20, 24, 58
shortcut 85
slave loop 8, 48, 70, 95–98
spinning (fibers) control 19
startup control 16
static mixer control 19, 36, 38
supervisory control 7
temperature control 14, 24, 27–29, 35, 38, 48, 52
web control 19, 24, 58
window of allowable gains 22, 47
Ziegler-Nichols 21, 48
tuning methods 1, 16, 47–62

U

update time of algorithm 105, 110–111
upset control (see also disturbance) 52–53, 55, 60, 92, 95–96, 99, 101, 105, 107

V

valve
actual position 57
actuator 66, 71
ball 66, 71
butterfly 12, 36, 66, 71
characteristic 12, 22, 31, 36–37, 71, 73, 95
column control 70
compressor control 66, 70
dead band 2, 19–20, 29, 33, 47, 48, 51, 53, 55, 60, 64, 71, 73
dead time 47, 53, 55–56, 58, 60, 71, 73
dynamic classes 66, 69
fail-open 4–6
level control 63, 70–71

    maintenance 64, 71–72
    pH control 70, 88
    positioner 64–71
    pressure control 66, 71
    resolution 27, 28, 29, 64, 71
    response time 66–69
    rotary 66, 70–71
    shaft design 66, 68, 71, 73
    size 11
    sliding stem (globe) 71
    specification 66
    split range 70
    stick-slip 31, 64, 102, 107
    temperature control 70–71
variability 1–2, 16, 19, 27, 107
variable
    controlled 11, 37, 90
    manipulated 3, 9, 11, 22, 28, 31, 36, 39–40
    process 2–3, 8–9, 11, 14, 16, 19, 27, 31, 36, 40, 51, 57–58, 60, 88, 90
variable speed drive (VSD) 3, 36
velocity 52, 125

W

waves
    pressure 99, 101
web control tuning 19, 24, 58
webs 147
window of allowable gains
    tuning 22, 47, 109–110

Z

Ziegler-Nichols
    tuning 21, 48